Air-Puff Tonometers

Challenges and insights

IOP Series in Medical and Biological Image Analysis

Series Editor: Robert Koprowski PhD DSc, University of Silesia, Katowice, Poland

Editor Bio: Robert Koprowski, PhD is a lecturer in the Department of Biomedical Computer Systems in Computer Science Institute at The University of Silesia, Katowice, Poland. His research interests include computer analysis and signal and biomedical image processing, as well as computer science technology in medicine and biotechnology. He is an author and co-author of 8 published books and has published over 140 papers.

Aims and Scope: The scope of this series involves the use of image analysis and processing methods in medicine and biology. New areas of medicine and biology in which image analysis and processing methods have not been used so far will be equally interesting to explore in this series - especially those operating automatically and repetitively in large patient populations. Additionally, any imaging method can be applied, starting with MRI through X-ray and ending with THV or other imaging methods that have not yet been invented. This series also includes works which are a combination of an engineer (algorithms for image analysis and processing) and a doctor (who verifies the practical clinical utility of the proposed image analysis and processing methods).

Air-Puff Tonometers

Challenges and insights

Robert Koprowski

University of Silesia, Katowice, Poland

IOP Publishing, Bristol, UK

ISBN   978-0-7503-2018-4 (ebook)
ISBN   978-0-7503-2016-0 (print)
ISBN   978-0-7503-2017-7 (mobi)

DOI   10.1088/2053-2563/aafee5

Version: 20190401

IOP Expanding Physics
ISSN 2053-2563 (online)
ISSN 2054-7315 (print)

British Library Cataloguing-in-Publication Data: A catalogue record for this book is available from the British Library.

Published by IOP Publishing, wholly owned by The Institute of Physics, London

IOP Publishing, Temple Circus, Temple Way, Bristol, BS1 6HG, UK

US Office: IOP Publishing, Inc., 190 North Independence Mall West, Suite 601, Philadelphia, PA 19106, USA

Contents

Preface

Intraocular pressure measurement methods that use air puffs are becomng more and more popular on the market. This type of measurement causes corneal deformation. This deformation is analyzed by an ultra-fast camera operating in visible or infrared light. It provides a sequence of images (videos) of the cornea being deformed in response to an air puff. Therefore, in addition to the pressure measurement, other corneal parameters are obtained—mainly related to the cornea's biomechanical properties. Biomechanical properties can be parameterized and further processed. This monograph deals with the latest achievements and methods related to the acquisition and practical suitability of biomechanical features and compares the results obtained from other devices that use other methods of ophthalmological diagnostics.

The book will definitely be of interest to scientists dealing with ophthalmology, as well as biomedical engineers or IT engineers. I encourage you to view the individual chapters.

Editor biography

Robert Koprowski

Robert Koprowski, MD (1997), PhD (2003), Habilitation (2015), is an employee of the University of Silesia in Poland, Institute of Computer Science, Department of Biomedical Computer Systems. For 20 years, he has been dealing with the analysis and processing of biomedical images with particular emphasis on the full automation of measurement for a large inter-individual variability of patients.

He is the author of dozens of papers with an impact factor (IF) and more than a hundred other papers, as well as the author or co-author of six books. Additionally, he is the author of several national and international patents in the field of biomedical devices and imaging. Since 2011, he has been a reviewer of grants and projects (including EU projects) in the field of biomedical engineering.

Contributor List

Magdalena Jędzierowska
University of Silesia, Poland

Michele Lanza
Università della Campania Luigi Vanvitelli, Italy

Marcelo Macedo
University of São Paulo, São Paulo, Brazil

Marcelo Hatanaka
University of São Paulo, São Paulo, Brazil

Hamed Momemi-Moghaddam Zahedan
University of Medical Sciences, Iran

Remo Susanna Jr
University of São Paulo, São Paulo, Brazil

Renato Ambrósio Jr
Federal University of the State of Rio de Janeiro (UNIRIO), Brazil

Hans R Vellara
University of Auckland, New Zealand

Akilesh Gokul
University of Auckland, New Zealand

Charles NJ McGhee
University of Auckland, New Zealand

Dipika Patel
University of Auckland, New Zealand

Pablo Pérez-Merino
Biomedical Research Institute Fundación Jiménez Díaz, Spain

Nandor Bekesi
Instituto de Óptica—CSIC, Spain

Antonio Fernández-López
Universidad Politécnica de Madrid, Spain

Sven Reisdorf
OCULUS Optikgeräte GmbH, Germany

IOP Publishing

Air-Puff Tonometers
Challenges and insights
Robert Koprowski

Chapter 1

Corvis ST tonometer and the possibility of analysing corneal deformation dynamics during intraocular pressure measurement

Magdalena Jędzierowska

1.1 Introduction

Nowadays, in medicine, a special role is played by early diagnostics, the task of which is to detect the first stages of disease, which can often be ambiguous and almost imperceptible. This is also the case in ophthalmology as doctors pay close attention to all parameters where the values even slightly deviate from the accepted norms during routine examinations and screening tests. New trends and ever-increasing needs are followed by the manufacturers of medical equipment, who offer new devices supporting the work of ophthalmologists. This trend can also be observed in tonometry tests, where non-contact tonometers, due to their minimally invasive nature, are most often used during these 'fast' routine examinations.

In 2010, the Corvis ST non-contact tonometer (OCULUS Optikgeräte GmbH, Wetzlar, Germany) was placed on the global market. Currently, it is the second device, just behind the ORA (ocular response analyser), which provides biomechanical parameters of the cornea. The measurements available owing to the applied solutions provide quite new indicators, previously unavailable in any other tonometer. The significance and impact of biomechanics of the eye (its geometry and material properties) on the processes responsible for correct vision had been known before [1]. However, it was only due to these new technologies that the possibility of *in vivo* measurements of biomechanical properties was developed, which in turn allowed one to study the influence of biomechanical parameters on refractive surgery, as well as their significance for the progression of ectatic diseases [2].

doi:10.1088/2053-2563/aafee5ch1

1.2 Basics of measurements with the Corvis ST tonometer

The Corvis ST tonometer is equipped with an ultra-high-speed Scheimpflug camera. It enables one to record a full corneal deformation video during intra-ocular pressure measurement. Tonometry and tachymetry are performed within the same measurement process. During the measurement lasting only 33 ms, the ultra-high-speed camera working at 4330 frames per second provides a series of 140 images [3]. The acquired images are cross-sections covering 8.5 mm of the cornea horizontally. It is worth noting that the air puff sent by Corvis ST is constant for every measurement and for each device. Its maximum value reaches 25 kPa. Blue LED light free of ultraviolet radiation with a length of 455 nm is used as a light source. Each of the images with a resolution of 640×480 pixels has 576 measurement points [3, 4].

During the measurement, the device increases the pressure of the air puff in proportion to the time. The shape of the cornea changes from convex to concave, passing through the characteristic stages: at the beginning of the measurement the cornea is in its natural convex shape, then, with increasing pressure of the applied air, the cornea reaches the first applanation state, i.e. flattening, when the air puff reaches its maximum, the cornea is at its highest concavity, then returning to its original shape, the cornea reaches the second applanation state, finally returning to its original shape, when the measurement process ends.

The described changes are detected on 140 Scheimpflug images. When applying the air puff, the blue LED light from the aperture illuminates the area from the front to the back surface of the cornea. Transparent cells of the cornea disperse the light in such a way that the examined area appears to shine. This is captured by the camera at an angle of 45° relative to the pupil, and the photosensitive area is also set at an angle of 45° with respect to the lens to achieve the same sharpness over the entire area. This setting provides sharp cross-sectional images of the cornea.

Intraocular pressure (IOP) is determined on the basis of the detected corneal applanation moments. Based on the Imbert–Fick principle, knowing the force needed to flatten the sphere as well as the surface of the flattening, it is possible to determine the pressure inside the sphere. In the analysed device, the pressure value is divided by the registered applanation surface.

1.3 Available parameters

After the measurement, the software provided with the Corvis ST tonometer displays a number of parameters related to corneal deformation. These parameters are based on a thorough analysis of the full process of the corneal dynamic response to the collimated air pulse that deforms it. The basic measurement of intraocular pressure (IOP) using the Corvis ST tonometer, its repeatability and accuracy have been the subject of many studies so far [5, 6]. However, the possibility of obtaining the already mentioned additional parameters describing corneal deformation dynamics determined the popularity of this device. Reproducibility of biomechanical parameters was confirmed for specific popula-tions [7], compared with other tonometers [8, 9], and good repeatability of selected parameters in the keratoconus examination was achieved [10]. In addition, the corneal dynamic parameters obtained from the Corvis ST tonometer were used repeatedly in the detection of glaucoma, keratoconus, as well as in the assessment of post-crosslinking patients.

In the first, original version of the software, the available biomechanical parameters allowed one to describe the characteristic moments of the deformation process, i.e. the moments of the first and second applanation (time of their occurrence, length), the moment of the highest corneal concavity, deformation amplitude (see table 1.1), as well as the graph showing changes in corneal velocity during the entire examination.

Table 1.1. Corvis ST—available parameters.

Parameter	Description
A1 length (mm)	Length of the flattened cornea at the first applanation
A1 velocity (mm ms^{-1})	Velocity of the corneal apex during the first applanation
A1 time (ms)	Time from the measurement beginning to the first applanation moment
A1 DeflAmp (mm)	Corneal deflection amplitude during the first applanation, determined as the displacement of the corneal apex in relation to the initial state without the whole eye movement.
DA ratio (2 mm)	Deformation amplitude ratio at 2 mm
DefA ratio (2 mm)	Deflection amplitude ratio at 2 mm
SP-A1	Stiffness parameter A1
SP-HC	Stiffness parameter HC
DA (mm)	Maximum deformation amplitude (measured at the moment of the highest corneal concavity). It is the actual sum of corneal deflection amplitude and whole eye movement
HC time (ms)	Time from the measurement beginning to the moment of reaching the highest concavity
HCDeflAmp (mm)	Corneal deflection amplitude at the moment of the highest corneal concavity
HCDeflArea (mm^2)	Highest concavity deflection area
HCdArclength (mm)	Highest concavity delta arc length
Peak distance (mm)	Distance between the corneal peaks at the moment of the highest corneal concavity
HC radius (mm)	Radius of corneal curvature during the moment of its highest concavity
InvRadMax (mm^{-1})	Maximum inverse radius
WEMmax (mm)	Maximum whole eye movement
A2 length (mm)	Length of the flattened cornea at the second applanation
A2 velocity (mm ms^{-1})	Velocity of the corneal apex during the second applanation
A2 time (ms)	Time from the measurement beginning to the second applanation moment

What is more, only one IOP value was available (no correction of this measurement was made), which in consequence did not allow the full use of the acquired dynamic parameters. It was only relatively recently, in 2016, that a set of new dynamic corneal deformation parameters was developed [2, 11] and an algorithm providing the so-called biomechanical-compensated IOP (bIOP) was presented [12]. The introduction of these new measurements in the form of comprehensive combinations allows for a more accurate and insightful analysis of corneal biomechanics during its deformation. It has been confirmed by the first works on the application of the introduced parameters, which, among others, have shown good keratoconus detection results [13].

New biomechanical parameters of dynamic corneal deformation have been included in the new version of Corvis ST tonometer software [14]. A tab allowing one to display the Vinciguerra Screening Report has been introduced as another additional element (figure 1.1). The aim is to compare the current measurement with normalised values.

Moreover, the Corvis biomechanical index (CBI) has been added, which enables one to separate healthy subjects (who are at low risk of corneal ectasia) from patients with increased ectasia risk (see figure 1.1). The unquestionable advantage of this report is a graphic representation of the available parameters: arc length, corneal velocity, deflection amplitude, deflection amplitude ratio (2 mm), deflection area, deformation amplitude, deformation amplitude ratio (2 mm), inverse concave radius, peak distance, whole eye movement, which are collected in table 1.1. In the new Corvis ST interface, there is also a tab known from previous versions that presents the dynamic corneal response (figure 1.2).

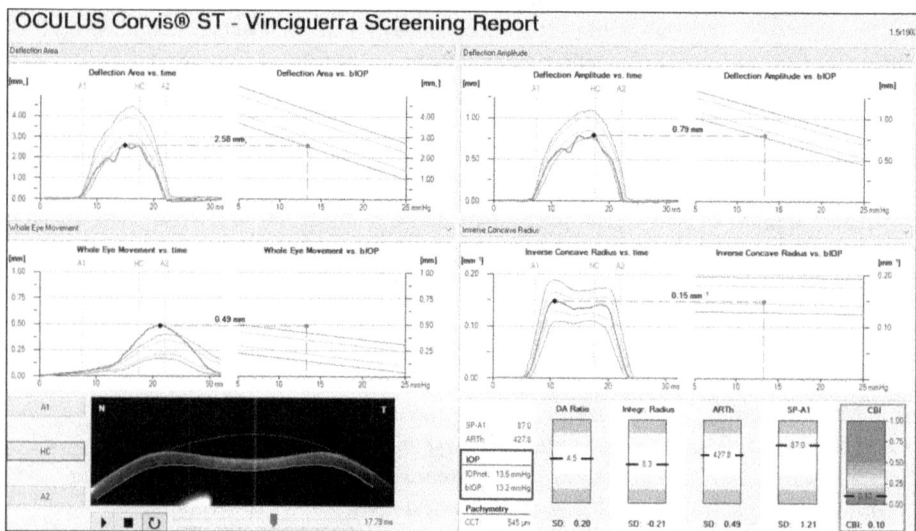

Figure 1.1. Screenshot of the Corvis ST display presenting the Vinciguerra Screening Report. The Corvis biomechanical index, which is the part of the report, is indicated by a red rectangle; the values of IOP (IOPnct, bIOP) delivered by Corvis ST are indicated by a blue rectangle.

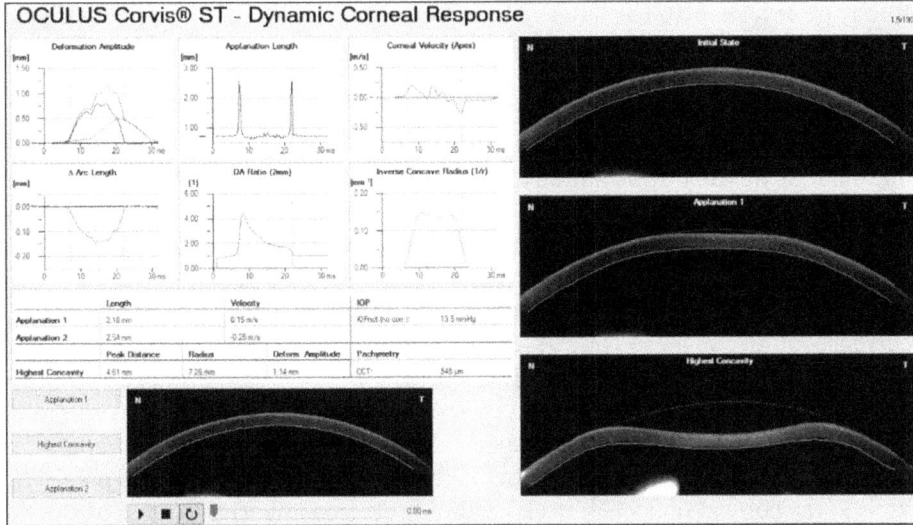

Figure 1.2. Screenshot of Corvis ST display presenting the dynamic corneal response tab.

1.4 Characteristics of dynamic corneal response parameters

All dynamic corneal response (DCR) parameters have been developed based on an insightful analysis of the individual stages of corneal deformation. Importantly, the parameters taking into account the movement of the eyeball (whole eye movement), which are described as deformation parameters, have been separated from parameters containing only 'pure' corneal displacement, which are described as deflection parameters.

The initial state, when the central corneal thickness (CCT) is measured, is followed by the first applanation phase, which is characterised by four parameters: length of the first applanation (A1 length), velocity of the corneal apex (A1 velocity), time of applanation event (A1 time) and corneal deflection amplitude (A1 DeflAmp). It should be noted here that there is a significant difference between the deflection amplitude and the already known parameter which is the deformation amplitude. The corneal deflection amplitude is defined as the displacement of the corneal apex in relation to the initial state without the whole eye movement (after the cornea from the initial state is applied onto the cornea at the time of deflection measurement). The deformation amplitude is a combination of both corneal displacement and whole eye movement.

Just after the first applanation, a new parameter is measured: deformation amplitude ratio at 2 mm (DA ratio (2 mm)). This is the ratio of the corneal apex deformation to the mean of deformation amplitudes measured on both sides 2 mm from the centre of the cornea. The next factor, namely deflection amplitude ratio at 2 mm (DefA ratio (2 mm)), is measured in the same way but it is devoid of the influence of the moving eyeball [15].

The stage when the cornea bends inwards until the maximum deformation is reached is associated with two further factors: stiffness parameter A1 (SP-A1) and stiffness parameter HC (SP-HC). The parameters discussed in [16] describe the pressure during the first applanation moment divided by the value of corneal displacement. For SP-A1, the displacement corresponds to the distance between the corneal apex at the beginning of the examination and the corneal apex at the time of the first applanation, whereas for SP-HC, it is the distance between the corneal apex at the time of the first applanation and the corneal apex at the time of maximum corneal deflection.

The next phase is the moment of the highest corneal concavity (HC), also called the oscillation phase in the literature, as the largest corneal oscillations are observed at this time. The following parameters are determined for this stage: time of occurrence of the highest corneal concavity (HC time), maximum deformation amplitude (DA), distance between two corneal peaks (peak distance), radius of corneal curvature at its highest concavity (HC radius), and maximum inverse radius (InvRadMax). InvRadMax is a parameter that indicates the maximum value of the inverse radius of the corneal curvature at the time of HC. In addition, parameters related to the corneal deflection itself (without eye movement) measured at the time of HC should be mentioned, i.e. corneal deflection amplitude (HCDeflAmp), corneal deflection area (HCDeflArea) and HC delta arc length (HCdArcLength). The last two are, respectively: the area between the outer edge of the cornea at the beginning of the examination and the outer edge of the cornea at the moment of its highest concavity (after subtraction of eye movement), and HCdArcLength is the difference in the arc length of the outer edge of the cornea between the initial state of the cornea and the time of HC, measured on each side, in the horizontal plane at a distance of 3.5 mm from the corneal apex [2].

When the air puff causing corneal deformation is switched off, the cornea returns to its initial state, passing through the second applanation phase. This stage is described by the following parameters: length of the second applanation (A2 length), time (A2 time) and velocity (A2 velocity) of the cornea during applanation. As previously mentioned, during measurements with the Corvis ST non-contact tonometer, the whole eye movement can be observed in addition to pure corneal deformation—the corneal displacement in the vertical plane is visible in the final state after the examination is over relative to its position before the examination. The maximum value of this displacement is described by the following parameter: maximum whole eye movement (WEMmax), measured during the return of the cornea to its natural shape, close to the second applanation moment [16, 17].

Interpretations as to the practical significance of the presented new dynamic corneal parameters as well as interpretations regarding their changes and values in terms of corneal biomechanics and the full deformation process during a tonometry test are presented consistently in recent works [2, 13, 16, 18–20].

1.5 Biomechanical-compensated IOP (bIOP) and dynamic corneal response parameters in the available literature

The dependence of intraocular pressure on age, central corneal thickness and its curvature has been indicated for a long time [21]. IOP values increase with age and for thicker corneas [22]. Before the technology of dynamic corneal deformation analysis had been disseminated to a wider group of scientists and ophthalmologists (thanks to the ORA and Corvis ST tonometer), a number of authors pointed out that these were not the only factors to be considered when correcting IOP values. Undoubtedly, biomechanical properties of the cornea, which can significantly influence the measurement error, also have an important role in this respect [23–28]. Currently, when research on corneal biomechanics—modelling its deformation in response to an external stimulus (air puff)—is so advanced and so many dependencies have been clarified, it is possible to use the new, corrected value of intraocular pressure developed by Joda *et al* [12]. Biomechanical-compensated IOP (bIOP), which is the new corrected pressure value, was developed taking into account the effect of variable corneal thickness as well as its material properties. The algorithm for determining the bIOP value was based on numerical simulations of dynamic corneal deformation. The model of the eyeball was developed using the finite element method, so it enables one to simulate the most important biomechanical properties of the cornea as well as the whole eye. It takes into account asphericity, variable corneal thickness, changes in corneal stiffness associated with age and many other parameters [12, 29]. The corrected pressure value available in the Corvis ST tonometer software is a modified version of the algorithm developed by Joda *et al* [2, 12].

1.5.1 Repeatability of biomechanical-compensated IOP (bIOP) values as well as new corneal parameters

It should be noted that the first publication on this topic was a collective work [30], in which the authors examined the repeatability and reproducibility of IOP, bIOP and seven selected dynamic corneal response (DCR) parameters. The study group included 32 patients. The obtained data indicated that most of the DCR parameters are characterised by good precision—coefficient of variation (CV) values were below 4% for repeatability and below 6% for reproducibility. For unadjusted IOP, CV values were 6.6% for repeatability and 7.6% for reproducibility. However, CV values for bIOP were slightly lower (a lower CV value indicates higher repeatability or reproducibility) −6.1% (repeatability) and 7.2% (reproducibility). A larger study group including 48 patients was collected in paper [31]. The repeatability analysis was performed for 35 parameters available in the Corvis ST tonometer, including IOP (but bIOP was not analysed). Repeatability of the corneal deformation parameters was investigated by analysing the intraclass correlation coefficient (ICC), while the repeatability coefficient was determined as the indicator of measurement reliability. The best results, where ICC $\geqslant$ 0.75, were obtained for 22 parameters. In turn, the ICC value dropped below 0.4 for only 6 parameters.

In other studies, the corrected intraocular pressure (bIOP) value was compared with the pressure measured using the Goldman applanation tonometer and the unadjusted IOP value measured with the Corvis ST tonometer [32]. The comparison was made for patients (69 eyes in total) before and after refractive procedures: SMILE (small-incision lenticule extraction) and LASIK (laser *in situ* keratomileusis). The final results indicated an insignificant correlation of the bIOP values with the central corneal thickness (CCT) both before and after the analysed procedures. However, the IOP values measured with the Goldman applanation tonometer, as already indicated in previous papers [33], were significantly dependent on changes in corneal thickness (statistical analysis according to Pearson: $P = 0.047$ for patients after LASIK, $P = 0.037$ for patients after SMILE). A significant correlation of $P = 0.05$ for patients after LASIK and $P = 0.003$ for patients after SMILE was also observed for the unadjusted IOP values measured with the Corvis ST tonometer.

The stability of the corrected pressure value (bIOP) after LASIK, and therefore lack of dependence of this parameter on the CCT value, was also indicated by Lee et al [19]. Before laser eye surgery the mean value of bIOP was 16.12 ± 1.66 mmHg, and after surgery 15.86 ± 1.32 mmHg (the mean for a group of 64 patients). For comparison, unadjusted IOP values before and after LASIK were 16.81 ± 1.87 mmHg and 14.19 ± 1.34 mmHg, respectively. The study of this team also confirmed the stability of this parameter after another refractive surgery—transepithelial photo-refractive keratectomy (PRK). It is worth noticing that the lack of significant differences in the bIOP values before and after transepithelial PRK had already been indicated in the earlier study of this research group [34]. In addition, in the work by Lee et al [19], the latest parameters of corneal deformation were analysed, namely: deformation amplitude ratio (2.0 mm) and integrated inverse radius (integrated total value of the inverse radius of the corneal curvature between the first and second applanation moments), which significantly increased after LASIK and PRK ($P < 0.001$—differences were significant when the P value was less than 0.05). This was in contrast to two other parameters: stiffness parameter at first applanation (SP-A1) and Ambrosio relational thickness to the horizontal profile (ARTh—a new parameter introduced as an addition to CCT measurement based on the corneal thickness profile in the temporal-nasal direction [13]), whose values after the performed procedure were significantly smaller ($P < 0.001$). One of the most important conclusions of the study was the indication of smaller changes in corneal biomechanics after the PRK procedure than after LASIK (resulting from smaller changes in the analysed DCR parameters).

1.5.2 Keratoconus

A significant part of the research literature focusing on the impact of corneal biomechanics on keratoconus should also be mentioned. In advanced form, it is easily recognisable in corneal topography, but difficult to detect in its early stages. Recent studies seem to confirm the long-standing hypothesis that when analysing the biomechanical parameters of the cornea, it is possible to detect keratoconus before the clinical symptoms of this disease are revealed [35–38]. One of the first parameters associated with keratoconus was deformation amplitude (DA), whose higher values (compared to the values obtained in patients without keratoconus) indicated the possibility of occurrence of keratoconus [37]. The time of the first applanation was also indicated as a parameter suitable for the diagnosis of the eyes with keratoconus [35]. Recent studies describe the effectiveness of the Corvis biomechanical index

(CBI), which was introduced to support the diagnosis of this disease. The first work on this topic that introduced CBI is the article by Vinciguerra *et al* [13]. In their study, this new parameter was evaluated in terms of its diagnostic capability to distinguish between healthy subjects and patients with keratoconus. For the purposes of this experiment, a large study group was collected, including two data sets containing 329 eyes each. Importantly, the data were from patients from different ethnic groups. The classification of patients suffering from keratoconus performed based on CBI showed high sensitivity (100%) and specificity (98.4%), the area under the ROC curve (AUC) was 0.990 for this parameter, whereas the percentage of correctly classified cases was 98.8%.

CBI was also studied for patients diagnosed with a subclinical form of keratoconus. Initial studies conducted for 12 cases indicated that despite the unchanged results in corneal tomography and topography, the Corvis biomechanical index value indicated the presence of corneal ectasia (CBI over 0.5) [39].

The results cited above indicate the potential of using CBI in clinical practice, which, together with a comprehensive assessment of corneal topography and tomography, would assist ophthalmologists in the diagnosis of keratoconus. However, it should be taken into account that despite the promising results, these are only the first studies in this respect and it is necessary to wait for the full confirmation of the reliability of this parameter.

1.6 Image processing for obtaining new biomechanical parameters of the cornea

Imaging with the ultra-high-speed Scheimpflug camera, which takes place in the non-contact Corvis ST tonometer, provides many possibilities for the analysis of the corneal deformation process. Advanced image processing and analysis tools are successfully used for this purpose. Using dedicated image processing methods, it is possible to analyse corneal vibrations during intraocular pressure measurement. As shown in various publications [14, 40–42], changes in corneal vibrations during its deformation can be successfully used as indicators or specific parameters allowing one to characterise the cornea. Corneal vibrations are so characteristic that presented in the form of properly developed parameters can allow for effective classification of, among others, patients with keratoconus. In an earlier paper [42], based on selected corneal vibration features, the authors recorded a specificity of 98%, sensitivity of 85% and accuracy of 82% in the automatic classification of healthy corneas and those with keratoconus.

The tools that the field of image processing has at its disposal remain an extremely valuable resource that can be used when analysing images from the Corvis ST tonometer. Currently, the biggest challenge, and at the same time difficulty, is the development of image analysis algorithms that will allow for automatic analysis, which are effective regardless of the intra-individual variability in patients.

1.7 Modelling of dynamic corneal deformation

Another area that is more and more often connected with corneal visualisation Scheimpflug technology (Corvis ST) is biomechanics. This field, related to the modelling and simulation of medical structures, uses the potential of the Corvis ST tonometer to develop a numerical model of the eyeball. The greatest difficulties and limitations are visible in the modelling of the eyeball in dynamic conditions, i.e. during a tonometry test. There are many requirements for a potential model. It should take into account the greatest possible number of parameters and factors that affect the reaction of the cornea as well as the entire anterior segment of the eye to an external stimulus. The ideal model should take into consideration variable corneal geometry, its shape and material properties (so as not to simplify them to one structure but to include all layers of the cornea and their properties). In addition, the boundary conditions of the model should take into account the influence of the limbus and accommodation system on deformation dynamics. An important aspect is also to consider the characteristics of fluid dynamics in the eye that differ significantly from that occurring under static conditions [43]. Moreover, the undoubted challenge is to include in the model corneal vibrations characteristic for tonometry tests.

As mentioned above, one of the difficulties in creating a biomechanical model of corneal deformation is the appropriate selection of its material parameters. In the literature, the cornea is described as a viscoelastic material characterised by so-called hysteresis [44], and this is the way it is most often modelled [43]. Recently, however, another concept prevails in the literature, according to which it is impossible to 'activate' the viscous properties of the cornea during tonometry tests, because the duration of the air puff causing corneal deformation is too short [44, 45].

1.8 Summary

The Corvis ST non-contact tonometer is one of the newest devices allowing the measurement of intraocular pressure and additional biomechanical parameters of the cornea. New parameters characterising corneal deformation, included in the so-called Vinciguerra Screening Report such as the Corvis biomechanical index, significantly extend the range of diagnostic capabilities of this equipment. Undoubtedly, it can also be said that owing to such a dynamic development of this technology, it will be possible to better understand the impact of biomechanics of the cornea, as well as the whole eye, on the genesis of eye diseases, and most importantly, it will allow for effective prevention and early detection of corneal diseases.

Acknowledgements

The author wishes to thank Professor Dorota Tarnawska from the Railway Hospital in Katowice and Wiesław Wrzosek from Oculus for valuable consultations and facilitating access to the Corvis ST tonometer.

References

[1] Śródka W 2009 Biomechanical model of human eyeball and its applications *Opt. Appl.* **39** 401–13

[2] Vinciguerra R, Elsheikh A, Roberts C J, Kang D S Y, Lopes B T, Morenghi E, Azzolini C and Vinciguerra P 2016 Influence of pachymetry and intraocular pressure on dynamic corneal response parameters in healthy patients *J. Refract. Surg.* **32** 550–61

[3] Ambrósio R Jr, Ramos I, Luz A, Correa Faria F, Steinmueller A, Krug M, Belin M W and Roberts C J 2013 Dynamic ultra high speed Scheimpflug imaging for assessing corneal biomechanical properties *Rev. Bras. Oftalmol.* **72** 99–102

[4] Hong J, Xu J, Wei A, Deng S X, Cui X, Yu X and Sun X 2013 A new tonometer—the Corvis ST tonometer: clinical comparison with noncontact and Goldmann applanation tonometers *Invest. Ophthalmol. Vis. Sci.* **54** 659–65

[5] Nemeth G, Hassan Z, Csutak A, Szalai E, Berta A and Modis L Jr 2013 Repeatability of ocular biomechanical data measurements with a Scheimpflug-based noncontact device on normal corneas *J. Refract. Surg.* **29** 558–63

[6] Hon Y and Lam A K C 2013 Corneal deformation measurement using Scheimpflug noncontact tonometry *Optom. Vis. Sci.* **90** e1–8

[7] Wu Y, Tian L and Huang Y 2016 *In vivo* corneal biomechanical properties with corneal visualization Scheimpflug technology in Chinese population *Biomed. Res. Int.* **2016** 9872643

[8] Lanza M, Iaccarino S, Cennamo M, Irregolare C, Romano V and Gironi Carnevale U A 2015 Comparison between Corvis and other tonometers in healthy eyes *Cont. Lens Anterior Eye* **38** 94–8

[9] Smedowski A, Weglarz B, Tarnawska D, Kaarniranta K and Wylegala E 2014 Comparison of three intraocular pressure measurement methods including biomechanical properties of the cornea *Invest. Ophthalmol. Vis. Sci.* **55** 666–73

[10] Ye C, Yu M, Lai G and Jhanji V 2015 Variability of corneal deformation response in normal and keratoconic eyes *Optom. Vis. Sci.* **92** e149–53

[11] Ambrósio R, Lopes B T, Faria-Correia F, Salomão M Q, Bühren J, Roberts C J, Elsheikh A, Vinciguerra R and Vinciguerra P 2017 Integration of Scheimpflug-based corneal tomography and biomechanical assessments for enhancing Ectasia detection *J. Refract. Surg.* **33** 434–43

[12] Joda A A, Shervin M M S, Kook D and Elsheikh A 2016 Development and validation of a correction equation for Corvis tonometry *Comput. Methods Biomech. Biomed. Eng.* **19** 943–53

[13] Vinciguerra R, Ambrósio R, Elsheikh A, Roberts C J, Lopes B, Morenghi E, Azzolini C and Vinciguerra P 2016 Detection of keratoconus with a new biomechanical index *J. Refract. Surg.* **32** 803–10

[14] Koprowski R and Wilczyński S 2018 Corneal vibrations during intraocular pressure measurement with an air-puff method *J. Healthc. Eng.* **2018** 5705749

[15] Leszczynska A, Moehler K, Spoerl E, Ramm L, Herber R, Pillunat L E and Terai N 2017 Measurement of orbital biomechanical properties in patients with thyroid orbitopathy using the dynamic Scheimpflug analyzer (Corvis ST) *Curr. Eye Res.* **43** 289–92

[16] Roberts C J, Mahmoud A M, Bons J P, Hossain A, Elsheikh A, Vinciguerra R, Vinciguerra P and Ambrósio R Jr 2017 Introduction of two novel stiffness parameters and interpretation of air puff-induced biomechanical deformation parameters with a dynamic Scheimpflug analyzer *J. Refract. Surg.* **33** 266–76

[17] Miki A, Maeda N, Ikuno Y, Asai T, Hara C and Nishida K 2017 Factors associated with corneal deformation responses measured with a dynamic Scheimpflug analyzer *Invest. Ophthalmol. Vis. Sci.* **58** 538–44

[18] Luz A, Faria-Correia F, Salomao M Q, Lopes B T and Ambrósio R Jr 2016 Corneal biomechanics: Where are we? *J. Curr. Ophthalmol.* **28** 97–8

[19] Lee H, Roberts C J, Kim T, Ambrósio R Jr, Elsheikh A and Yong Kang D S 2017 Changes in biomechanically corrected intraocular pressure and dynamic corneal response parameters before and after transepithelial photorefractive keratectomy and femtosecond laser-assisted laser *in situ* keratomileusis *J. Cartaract Refract. Surg.* **43** 1495–503

[20] Vinciguerra R, Romano V, Arbabi E M, Brunner M, Willoughby C E, Batterbury M and Kaye S B 2017 *In vivo* early corneal biomechanical changes after corneal cross-linking in patients with progressive keratoconus *J. Refract. Surg.* **33** 840–6

[21] Doughty M J and Zaman M L Human corneal thickness and its impact on intraocular pressure measures: a review and meta-analysis approach *Surv. Ophthalmol.* **44** 367–408

[22] Kotecha A, White E T, Shewry J M and Garway-Heath D F 2005 The relative effects of corneal thickness and age on Goldmann applanation tonometry and dynamic contour tonometry *Br. J. Ophthalmol.* **89** 1572–5

[23] Ambrosio R, Nogueira L P, Caldas D L, Fontes B, Luz A, Cazal J, Alves M and Belin M 2011 Evaluation of corneal shape and biomechanics before LASIK *Int. Ophthalmol. Clin.* **51** 11–39

[24] Ogbuehi K C and Osuagwu U L 2013 Corneal biomechanical properties: Precision and influence on tonometry *Cont. Lens Anterior Eye* **37** 124–31

[25] Liu J and Roberts C J 2005 Influence of corneal biomechanical properties on intraocular pressure measurement: quantitative analysis *J. Cataract Refract. Surg.* **31** 146–55

[26] Kotecha A 2007 What biomechanical properties of the cornea are relevant for the clinician? *Surv. Ophthalmol.* **52** S109–14

[27] Luce D A 2005 Determining *in vivo* biomechanical properties of the cornea with an ocular response analyzer *J. Cataract Refract. Surg.* **31** 156–62

[28] Franco S and Lira M 2009 Biomechanical properties of the cornea measured by the ocular response analyzer and their association with intraocular pressure and the central corneal curvature *Clin. Exp. Optom.* **92** 469–75

[29] Elsheikh A 2010 Finite element modeling of corneal biomechanical behavior *J. Refract. Surg.* **26** 289–300

[30] Lopes B T, Roberts C J, Elsheikh A, Vinciguerra R, Vinciguerra P, Reisdorf S, Berger S, Koprowski R and Ambrósio R Jr 2017 Repeatability and reproducibility of intraocular pressure and dynamic corneal response parameters assessed by the Corvis ST *J. Ophthalmol.* **2017** 8515742

[31] Miki A, Maeda N, Asai T, Ikuno Y and Nishida K 2017 Measurement repeatability of the dynamic Scheimpflug analyzer *Jpn. J. Ophthalmol.* **61** 433–40

[32] Chen K-J, Joda A, Vinciguerra R, Eliasy A, Mohi Sefat S M, Kook D, Geraghty B, Roberts C J and Elsheikh A 2018 Clinical evaluation of a new correction algorithm for dynamic Scheimpflug analyzer tonometry before and after laser *in situ* keratomileusis and small-incision lenticule extraction *J. Cataract Refract. Surg.* **44** 581–8

[33] Tonnu P, Ho T, Sharma K, White E, Bunce C and Garway-Heath D 2005 A comparison of four methods of tonometry: method agreement and interobserver variability *Br. J. Ophthalmol.* **89** 847–50

[34] Lee H, Roberts C J, Ambrósio R Jr, Elsheikh A, Kang D S Y and Kim T-i 2017 Effect of accelerated corneal crosslinking combined with transepithelial photorefractive keratectomy on dynamic corneal response parameters and biomechanically corrected intraocular pressure measured with a dynamic Scheimpflug analyzer in healthy myopic patients *J. Cataract Refract. Surg.* **43** 937–45

[35] Elham R, Jafarzadehpur E, Hashemi H, Amanzadeh K, Shokrollahzadeh F, Yekta A and Khabazkhoob M 2017 Keratoconus diagnosis using Corvis ST measured biomechanical parameters *J. Curr. Ophthalmol.* **29** 175–81

[36] Vellara H R and Patel D V 2015 Biomechanical properties of the keratoconic cornea: a review *Clin. Exp. Optom.* **98** 31–8

[37] Ali N Q, Patel D V and McGhee C N J 2014 Biomechanical responses of healthy and keratoconic corneas measured using a noncontact Scheimpflug-based tonometer *Invest. Ophthalmol. Vis. Sci.* **55** 3651–9

[38] Roy A S, Shetty R and Kummelil M K 2013 Keratoconus: a biomechanical perspective on loss of corneal stiffness *Indian J. Ophthalmol.* **61** 392–3

[39] Vinciguerra R, Ambrósio R, Roberts C J, Azzolini C and Vinciguerra P 2017 Biomechanical characterization of subclinical keratoconus without topographic or tomographic abnormalities *J. Refract. Surg.* **33** 399–407

[40] Koprowski R, Ambrósio R and Reisdorf S 2015 Scheimpflug camera in the quantitative assessment of reproducibility of high-speed corneal deformation during intraocular pressure measurement *J. Biophotonics* **8** 968–78

[41] Koprowski R 2015 Open source software for the analysis of corneal deformation parameters on the images from the Corvis tonometer *Biomed. Eng. Online* **14** 31

[42] Koprowski R and Ambrósio R 2015 Quantitative assessment of corneal vibrations during intraocular pressure measurement with the air-puff method in patients with keratoconus *Comput. Biol. Med.* **66** 170–8

[43] Kling S, Bekesi N, Dorronsoro C, Pascual D and Marcos S 2014 Corneal viscoelastic properties from finite-element analysis of *in vivo* air-puff deformation *PLoS One* **9** e0104904

[44] Sinha Roy A, Kurian M, Matalia H and Shetty R 2015 Air-puff associated quantification of non-linear biomechanical properties of the human cornea *in vivo J. Mech. Behav. Biomed. Mater.* **48** 173–82

[45] Simonini I, Angelillo M and Pandolfi A 2016 Theoretical and numerical analysis of the corneal air puff test *J. Mech. Phys. Solids* **93** 118–34

Chapter 2

Air-puff devices, not just tonometry

Michele Lanza

2.1 Introduction

Intraocular pressure (IOP) evaluation is a crucial phase of a routine eye visit, more in glaucoma-affected patients, for which elevated IOP is the only risk factor that physicians are able to modify [1]. This is the reason for the importance of IOP measurement: it is an important element of glaucoma diagnosis and management [2]. Goldmann applanation tonometry (GAT) represents the 'gold standard' method for IOP evaluation [3]. However, many elements may affect its precision, among these, there are those related to the morphology of the eye, such as central corneal thickness (CCT) or corneal curvature and those related to corneal biomechanical properties. CCT has been demonstrated to bias IOP measurements by GAT, inducing IOP underestimation in thin corneas and overestimation in thick ones [4].

The ocular response analyzer (ORA) (Reichert Ophthalmic Instrument, Depew, NY, USA) is a kind of air-puff tonometer, first proposed more than 10 years ago, allowing IOP evaluation to take into consideration corneal biomechanical properties [5]. Analysis of corneal deformation observed while an air puff of the ORA is modifying the corneal surface provides two new parameters: corneal hysteresis (CH) and corneal resistance factor (CRF), which were also used to calculate another kind of IOP measurement [5]. Many papers have evaluated the influence in healthy eyes, in eyes affected by ocular diseases or eyes that have previously undergone ocular surgery [6–22], obtaining useful information for ophthalmic clinical practice [10, 15, 18, 21] and surgical screening [9–11, 17, 19]. Corvis ST (CST, Oculus, Wetzlar, Germany) is an air-puff device purposed to measure IOP and to evaluate corneal deformation during measurements due to an ultra-high-speed Scheimpflug camera that records the corneal changes in 4330 frames/sec along an 8 mm horizontal corneal coverage, while an air-puff indentation causes corneal deformation [23–26]. Comparing the parameters provided both by ORA and CST did not produce comparable results because the principle of working and the analysis of data of the two instruments are very different [27–29]. It is highly probable that the most important difference in the results obtained by the two devices is caused by the air puff applied during measurements. ORA provides one with no fixed force whereas CST uses a fixed force [1, 14, 17, 26–29].

Fields of applications of *in vivo* corneal biomechanical properties' analysis could be pressure evaluation of the eye but also improvement in refractive surgery screening, in order to avoid corneal degenerations such as corneal ectasia, and earlier diagnosis of corneal diseases and improvement in their management.

Keratoconus (KC) is a corneal degeneration where it is possible to detect a progressive thinning and increasing of both anterior and posterior corneal curvature [30]. The only treatment available for this disease until some years ago was the corneal transplant [31], but nowadays KC progression can be stopped by using a new kind of treatment: corneal cross-linking (CXL) [32, 33]. It has been stated that this kind of therapy is more effective if performed at earlier stages [32, 33]; because of this, it is very important that physicians have the tools to make KC diagnosis as early as possible. Analysis of corneal biomechanical properties could unveil diagnostic characteristics of KC eyes that could help in recognizing them in the early stages. CXL treatment has changed in many aspects from the first protocol to the more recent application [32, 33]; more details about corneal changes in KC eyes could allow one to design new profiles of treatment and new protocols in order to improve the customization of therapies, so improving reliability and effectiveness.

Correction or refractive defects using an excimer laser to modify the corneal profile with laser *in situ* keratomileusis (LASIK), in association with femtosecond laser assisted procedures, and photorefractive keratectomy (PRK), is nowadays one of the most diffused worldwide surgical procedures in the world [34–37], and this topic is so important that new techniques are continually designed in order to obtain better and more stable results [34–37]. Complications of this kind of treatment are dramatically reduced from the first step in this field today, due to improvement in every aspect of the procedures. However, some problems are still occurring, such as corneal ectasia or incorrect refractive corrections due to errors in treatment or screening [38–47].

2.2 How the instruments work

2.2.1 ORA

The ORA produces a corneal deformation via a rapid air pulse. Corneal shape changes are detected by a sophisticated electro-optical system that records the starting shape, the first applanation, the inward shape modification and the second applanation; moreover, it is able to measure two applanation IOP measurements—one while the cornea moves inward and the other as the cornea moves outward [5]. The resistance that the cornea offers to the air puff determines the difference in time and in extension of inward and outward applanation. The instruments record the first IOP as P1 and the second IOP as P2; the average of both measurements provides the so-called Goldmann-correlated IOP (IOPg). When discussing corneal biomechanical properties, one should note that the difference between these two IOP measurements will produce a parameter called corneal hysteresis (CH), so $CH = P1 - P2$. CH could be defined as the result of the viscous damping within corneal tissues. It is determined and influenced by the viscosity of glycosaminoglycans (GAGs) and proteoglycans (PGs), as well as by a collagen matrix interaction [48]. Because of CH, ORA is able to present two other parameters: corneal-compensated IOP (IOPcc) and a corneal resistance factor (CRF). IOPcc is a value theoretically less influenced by the morphological corneal properties than other IOP measuring devices. CRF is considered to be a parameter indicating the corneal resistance to external modifications [49] and it is calculated by this formula: $CRF = k1 \times (P1 - 0.7 \times P2) + k2$ ($k1$ and $k2$: constants).

2.2.2 Corvis ST

The device automatically starts ejecting the air puff when the corneal apex is aligned, and this is checked by a Purkinje corneal reflex. The images of corneal deformation are acquired by a high definition Scheimpflug camera with a blue light LED with UV-free 455 nm, covering 8.5 mm horizontally of a single slit. The UHS Scheimpflug camera is capable of taking 4300 frames per second. The recording measurement time is 31ms, which allows one to acquire 140 digital frames. Each image has 576 measuring points across each corneal surface, with a resolution of 200×576 pixels. Activation of the camera starts just before the air pump emission. Thus, it is possible to analyze changes in corneal shape from an undisturbed position to the ones during air-puff action. A quality score (QS) is used for the evaluation of the quality of images and measurements provided by scans [50].

The immediate corneal reaction to the air-puff action is an increase in the inward change of curvature from a convex shape to an almost plane (applanation); afterwards, the shape modification continues, resulting in a concave corneal profile. Hereafter, when the pressing effect of the air puff is completed, the cornea returns to the original shape. In a while, another applanation profile is assumed before obtaining the natural curvature. The camera is able to record every time, the distance of the corneal layers and IOP values related to every frame acquired. Afterwards, the software will present all these data and provide the characteristic Corvis parameters [23].

The second version of the Corvis ST software (version 1.00b32 revision 792) also provides the corneal thinnest point and an IOP value presented with CCT compensating algorithms.

2.3 Accuracy and repeatability

2.3.1 ORA

Studies about ORA reliability showed a very good intra- and inter-observer repeatability; also, if the score of measurement is lower than 3.5 it should always be ignored [51–53].

2.3.2 Corvis

Authors who evaluated the reliability and repeatability of corneal deformation parameters provided by CST did not show unanimous results [24, 54–56]. It is important to remember that devices are not always utilized with the same software. The first one suggested that only HCDA and AT1 have good repeatability whereas the later ones detected a less than 10% variation coefficient in the following parameters provided by CST: IOP, CCT, AT1, AL1, HCT, HCR, HCDA, HCPD; HCDT and AT2.

More recent papers confirm these results [57, 58], suggesting that the last version of the software improved the reliability of the device.

2.4 Corneal deformation in healthy corneas

2.4.1 ORA

In healthy corneas, a significant correlation has been detected between corneal thickness and biomechanical parameters provided by ORA, CH and CRF [48, 59–70]. Thicker corneas, indeed, show more opposition to deformation induced by external forces and higher damping ability. For this reason, CH is negatively correlated with IOP, because the higher the corneal tension, the faster the original position is restored after external forces have deformed the cornea [59, 67, 71, 72]. Therefore, it is easy to understand that CRF is positively correlated with IOP because they work together in producing a corneal resistance to the external forces aiming to deform it [73].

Because of the natural changes occurring to corneal structural and morphological characteristics, variations in CH and CRF have been described in relation to age and, in particular, a reduction of both parameters has been detected with increasing age [72]. Evaluating the relation between corneal curvature and CH, a significant negative one has been detected with lower CH values observed in flattened corneas [73].

The effects of refractive defect, gender, and axial length on ORA measurements have been demonstrated as not significant [59, 74, 75].

2.4.2 Corvis

According to the data observed by Huseynova *et al* [28], in more than one thousand eyes screened both with ORA and CST, before undergoing corneal excimer laser refractive surgery, the HCDA values seem to be more related to IOP than to CCT, and HCRC is more influenced by pachimetry than by IOP.

A correlation between IOP and CCT, AT1, AL1 and HCDA values provided by CST has been detected by Valbon *et al* [76] evaluating only one eye of healthy subjects.

Analyzing 29 eyes of healthy subjects, Bak-Nielsen *et al* [55] evaluated the influence of age on CST measurements. Older subjects showed higher HCDA and HCPD, suggesting that cornea deformability increases with age. It is important to highlight that some papers have suggested that corneal deformation is decreased in older subjects when analyzing ORA data [18, 22].

No significant correlations between corneal deformation parameters measured by CST and morphological ones such as corneal volume, spherical equivalent corneal curvature and CCT were found by Lanza *et al* [77] in 76 eyes of 76 healthy subjects.

The variations of the results observed in the published papers about this topic could rely on the differences in the study population analyzed.

2.5 Corneal deformation in diseased corneas

2.5.1 ORA

Biomechanical parameters measured by ORA are both reduced in keratoconus (KC) eyes' severity [78–82]. This could be explained by the weakness of these corneas due to the inner reduction of the strength of ties among corneal fibrillae and with corneal thinning.

Papers evaluating the influence of diabetes on CH and CRF measurements do not show unanimous results, mostly because of diabetes classification [83–86]. The eyes of patients with high HbA1c showed higher values of CH and CRF, suggesting an increase in the viscosity of the corneal tissue, probably due to glycosylation of both PGs and GAGs [87, 88]. In cases of corneal edema or corneal swelling, due to inflammation or after eye surgery, CH and CRF showed lower values [89, 90].

Evaluating glaucoma eyes, CH has been shown to be a predictive parameter for visual field worsening [91–94]. Even after reduction of IOP in glaucoma eyes, CH is still lower than in healthy eyes, suggesting CH could represent an independent parameter to evaluate glaucoma progression [95, 96]. Thus, an increase of CH values could represent a good compensation of IOP in these patients, suggesting a halt in glaucoma progression [7].

2.5.2 Corvis

According to results observed in the study published by Tian *et al* [97], HCDA could be considered to be a sensitive parameter in KC screening. They compared corneal deformation parameters measured in 60 healthy and in 60 KC eyes. Similar results were observed in the paper by Ali *et al* [98], and HCDA was considered to be a useful parameter in the diagnosis and management of KC eyes.

Lanza *et al* [99] evaluated CST parameters obtained scanning healthy subjects, eyes with a history of previous PRK, KC eyes and eyes with KC that underwent CXL. According to the results observed, in diseased corneas, it seems that corneal curvature is able to affect corneal deformation more than corneal thickness. Analysis of the other groups of subjects included in the study provided other interesting results: AT1 showed values negatively correlated to KM in healthy, KC and post-CXL corneas. This suggests that it is harder to applanate flatter corneas in these groups whereas in eyes that underwent PRK it seemed easier.

Perez-Rico *et al* [100] analyzed type II diabetes mellitus (DM) patients and healthy subjects. The CST parameters did not show significant variation in uncontrolled DM patients, controlled DM ones and healthy subjects.

Ye *et al* [58] evaluated 12 healthy and 12 KC eyes, observing a higher HCDA and a HCRC in patients affected by KC.

Analyzing 60 glaucoma eyes, Lee *et al* [101], analyzed 61 healthy eyes and 60 glaucoma eyes observing higher AV2 and HC peak distance values and lower HCT in glaucoma eyes.

KC was the disease most studied in the published papers, providing information about CST use in clinical practice [58, 97, 98], and more information about the changes of corneal deformation in diseased eyes [99]. It is important to highlight that corneal deformation analysis is not only related to diseased eyes affected by KC [100, 101].

2.6 Corneal deformation after surgery

2.6.1 ORA

Laser excimer refractive surgery, with both PRK and LASIK techniques, determines a corneal ablation that leads to corneal weakening because of the loss of corneal lamellae cohesion [8, 102–105]. These changes are reflected in CH and CRF measurement that appear to be lower after refractive surgery [8, 9, 22]. Usually, modifications in corneal structure and biomechanical properties appeared stable over a six-month follow-up [22, 106]. Analysis of biomechanical properties before refractive surgery could help in better screening eyes in order to avoid ectasia after these procedures [11]. Analyzing the effect of other types of corneal surgery techniques on CH and CRF, intrastromal corneal ring implants did not induce CH and CRF variations; even if, at a twenty-month follow-up there is a difference in ORA parameters [78, 107]. Twelve months after penetrating keratoplasty (PK) no significant difference in CH and CRF compared to healthy eyes has been detected [108]. Moreover, the same results showed a comparison between eyes that underwent PK and those that underwent deep anterior lamellar keratoplasty after more than two years from surgery [109].

2.6.2 Corvis

Comparing the data obtained scanning different kinds of eyes with CST such as KC eyes, KC eyes previously undergone CXL and healthy eyes, Bak-Nielsen et al [54] observed differences in some CST parameters between healthy and KC eyes, whereas comparison between KC eyes and those with KC after CXL did not show significant discrepancies.

No significant variations have been detected comparing CST parameters before and after both conventional CXL and accelerated one in KC affected eyes by Tomita et al [110], suggesting stability of the corneas and the safety of both therapies.

According to the results proposed by Maeda et al [111], HCDA resulted to be higher in eyes affected by KC that underwent penetrating keratoplasty compared to values measured in healthy eyes and in those that underwent Descemet stripping automated endothelial keratoplasty, whereas they were very close to the those that underwent deep anterior lamellar keratoplasty. HCRC showed higher values in healthy eyes compared to others that underwent corneal surgery.

Chen et al [57] analyzed healthy eyes and those submitted to myopic PRK; AT1, AV1, HCRC provided by CST, showed higher values in healthy subjects whereas AV2 and DA showed higher values in PRK subjects.

Pedersen *et al* [56] evaluated 31 healthy subjects with 95 patients that, one year before, underwent refractive surgery (different kinds of procedures), and particularly analyzed CST parameters. The results observed suggest that corneal deformation is higher in more invasive surgical techniques.

Both of the last papers mentioned suggested interesting theories that need to be confirmed in more organized studies.

The consequences of small incision lenticule extraction (SMILE) on corneal deformation characteristics were analyzed both by Shen *et al* [112] and Mastropasqua *et al* [113] using CST. The first paper observed significant variations in AT1, AT2 and HCDA and concluded by suggesting that lenticule extraction influences corneal deformation changes more than formation. The second paper provides results stating that CST parameters such as HCDA, AT1 and AT2 decrease after seven days from surgery and these changes are stable at a three-month follow-up.

According to the results observed by Hassan *et al* [114]—who analyzed the effects both of both PRK and LASIK on corneal deformation properties measured with CST—no significant changes in CST parameters have been observed after one month of surgery. Of course, these are preliminary studies and more in-depth work, analyzing longer follow-up and specific evaluation of changes in relation to the refractive defects treated are needed.

2.7 Studies regarding software improvements

As previously described, air-puff tonometers have provided a new way to analyze the cornea: evaluating the biomechanical properties. ORA has been the first instrument to enable this new kind of study. It has been improved and new developments have been proposed from the first introduction to the present day, but software implementations are missing from some years.

CST is a device that has aroused much interest not just in the scientific ophthalmologist community but also in the field of engineering. Many published papers suggest software improvements for these instruments in order to improve their efficacy and accuracy [115–121]. The common purpose of these studies is to develop new algorithms or new tools that are able to improve the corneal evaluation abilities of CST [115–121]. Koprowski *et al* [115] analyzed images acquired by CST in normal subjects and developed an algorithm providing evaluation of corneal morphological changes during air-puff force on a frequency range from 150 Hz to 500 Hz. Using this new software, the authors proposed a classification of three stages of the eyeball deflection when a force is applied to the eye itself. The software and the classification could be applied to the screening and management of corneal diseases. One of the earlier characteristics of diseases such as corneal ectasia or keratoconus is corneal weakness and this could be more easily detected with this last tool. The same group of authors proposed three new parameters: (a) the degree of corneal reaction relative to static position, (b) corneal length changes, and (c) ratio of amplitude changes to the corneal deformation length, in a later paper [116]. More papers, by the same study team, proposed more improvements in the corneal deformation analysis, analyzing also ocular structures such as the scleral layer, ocular muscles, and moreover other parameters such as axial length changes and absolute corneal displacement during a CST exam due to air-puff force [117, 118]. This different approach in corneal deformation evaluation, taking into account more parameters also not strictly related to the cornea itself, opens a new way of studying corneal response to different stimuli and, in general, to interactions among ocular structures. Another enhancement in this field could be the overall comprehension of corneal vibration during a corneal scan with CST with high frequency (>100 Hz) analysis. It could allow one to rely on more parameters able to improve the efficacy of KC screening [118]. Because of improvements obtained from previous studies [115–117], Koprowski proposed a new open software that is able to provide ten new parameters to enhance corneal deformation study [119] due to a new kind of approach in image analysis of video frames acquired by Corvis using high frequency (>100 Hz) and to a more precise definition of the cornel-scleral limit definition enabled by new software specially developed for this purpose. The new proposed parameters are: (1) corneal length changes, defined as the corneal length calculated with high frequency analysis; (2) ratio–amplitude changes/corneal deformation length, defined as the ratio between amplitude and length of corneal deformation calculated with high frequency analysis; (3) ratio–corneal reaction/corneal static position, defined as the ratio between the corneal morphological changes and the situation before air puff starts; (4) maximum amplitude of the eyeball reaction, defined as the measurement of the overall eyeball deformation to air puff; (5) maximum amplitude for the frequency >100 Hz, defined as the maximum corneal morphological deformation measured with high frequency analysis of Corvis video/images; (6) time cornea deformations >100 Hz, defined as the time needed to reach the maximum corneal morphological deformation measured with high frequency analysis of Corvis video/images; (7) automatic distinction between the left and right eye; (8) asymmetry in the work of muscles—left or right eye, defined as the differences in extraocular muscle reaction to air puff; (9) absolute of the cornea reaction, defined as the corneal deformation to air puff without the other eyeball components' influence; (10) scleral reaction asymmetry, defined as the difference of scleral deformations between both eyes due to air puff due to scleral borders' definition by new software.

Innovations in defining these new parameters are related to (1) above 100 Hz frequency analysis of video frames acquired by CST (parameters 5 and 6); automatic detection of right/left eye due to analysis of the extraocular muscles' response and the related effect on the scleral layer (parameters 7 and 8).

Papers confirming the clinical usefulness of these improvements are scarce. One of the main reasons is that CST is now considered to be a research tool and it is not so widely used. Moreover, evaluation of the new software would cause a loss for the manufacturer of the device and this increases the cost for a unit that one would want to compare them both.

A new approach in CST frames' analysis has been proposed by Ji *et al* [120]. They applied a segmentation strategy to improve the quality of the images and afterwards apply a curvature algorithm.

Tian *et al* [121] introduced a new method to present images acquired by CST during scans, introducing a custom-designed program: PolyU. This new approach provided parameters previously present in manufactured software, like HCDA, and some new ones that could be used in the diagnosis of corneal disease. The new parameters are: deformation highest concavity time (DHC-time), defined as the time from corneal deformation starting to the highest concavity; maximum corneal inward velocity ($V_{in,max}$), defined as the maximum corneal inward deformation velocity at the centreline; maximum corneal outward velocity ($V_{out,max}$), defined as the maximum corneal outward deformation velocity at the centreline; corneal creep rate (V_{creep}), defined as the gradient of corneal deformation against time after the end of the air puff; corneal contour deformation (CCD), defined as the distance between the original contour and the contour at the highest concavity; maximum deformation area (MA), defined as the maximum corneal deformation area within the two knees; maximum deformation area time (MA-time), defined as the time from deformation starting to the maximum deformation area occurred; temporal symmetry factors (T_{sym}), defined as the ratio of loading and unloading areas under the corneal deformation against the time curve. Two groups of eyes, one ketaconus the other healthy, were compared and the differences among these new parameters were evaluated.

2.8 Conclusions

Improvement in technology has enabled physicians to solve many medical riddles and to improve the quality of life of their patients and of their work. Unfortunately, there is still much to learn; for example, there is the relation between corneal morphology and corneal deformation [122–128]. In the past, many papers evaluated only the first characteristics because the second characteristics were impossible to study *in vivo*; but now it is possible using different devices and interesting results have been provided with more to come. Analysis of corneal deformation and the relationship with the structural properties of the corneas could suggest a better way to screen and manage both healthy subjects and diseased ones. Moreover, it is possible to analyze information to obtain more accurate measurement of IOP and provide better screening for patients with glaucoma because current devices are biased by morphological characteristics.

References

[1] Coleman A L and Miglior S 2008 Risk factors for glaucoma onset and progression *Surv. Ophthalmol.* **53** S3–10

[2] Arora R, Bellamy H and Austin M 2014 Applanation tonometry: a comparison of the Perkins handheld and Goldmann slit lamp-mounted methods *Clin Ophthalmol.* **26** 605–10

[3] Standardization IOf 2001 *Ophthalmic Instruments—Tonometers: ISO8612* (Geneva, Switzerland: International Organization for Standardization)

[4] Wang J, Cayer M M, Descovich D, Kamdeu-Fansi A, Harasymowycz P J, Li G and Lesk M R 2011 Assessment of factors affecting the difference in intraocular pressure

measurements between dynamic contour tonometry and goldmann applanation tonometry *J. Glaucoma* **20** 482–7

[5] Luce D A 2005 Determining *in vivo* biomechanical properties of the cornea with an ocular response analyzer *J. Cataract Refract. Surg.* **31** 156–62

[6] Martinez-de-la-Casa J M, Garcia-Feijoo J, Fernandez-Vidal A, Mendez-Hernandez C and Garcia-Sanchez J 2006 Ocular response analyzer versus Goldmann applanation tonometry for intraocular pressure measurements *Invest. Ophthalmol. Vis. Sci.* **47** 4410–4

[7] Congdon N G, Broman A T, Bandeen-Roche K, Grover D and Quigley H A 2006 Central corneal thickness and corneal hysteresis associated with glaucoma damage *Am. J. Ophthalmol.* **141** 868–75

[8] Pepose J S, Feigenbaum S K, Qazi M A, Sanderson J P and Roberts C J 2007 Changes in corneal biomechanics and intraocular pressure following LASIK using static, dynamic, and noncontact tonometry *Am. J. Ophthalmol.* **143** 39–47

[9] Ortiz D, Piñero D, Shabayek M H, Arnalich-Montiel F and Alió J L 2007 Corneal biomechanical properties in normal, post-laser *in situ* keratomileusis, and keratoconic eyes *J. Cataract Refract. Surg.* **33** 1371–5

[10] Kotecha A 2007 What biomechanical properties of the cornea are relevant for the clinician? *Surv. Ophthalmol.* **52** S109–14

[11] Kerautret J, Colin J, Touboul D and Roberts C 2008 Biomechanical characteristics of the ectatic cornea *J. Cataract Refract. Surg.* **34** 510–3

[12] Fontes B M, Ambrósio R Jr, Alonso R S, Jardim D, Velarde G C and Nosé W 2008 Corneal biomechanical metrics in eyes with refraction of −19.00 to +9.00 D in healthy Brazilian patients *J. Refract. Surg.* **24** 941–5

[13] Rosa N, Lanza M, Borrelli M, Palladino A, Di Gregorio M G and Politano L 2009 Intraocular pressure and corneal biomechanical properties in patients with myotonic dystrophy *Ophthalmology* **116** 231–4

[14] Narayanaswamy A, Chung R S, Wu R Y, Park J, Wong W L, Saw S M, Wong T Y and Aung T 2011 Determinants of corneal biomechanical properties in an adult Chinese population *Ophthalmology* **118** 1253–9

[15] Terai N, Raiskup F, Haustein M, Pillunat L E and Spoerl E 2012 Identification of biomechanical properties of the cornea: the ocular response analyzer *Curr. Eye Res.* **37** 553–62

[16] Kara N, Altinkaynak H, Baz O and Goker Y 2013 Biomechanical evaluation of cornea in topographically normal relatives of patients with keratoconus *Cornea* **32** 262–6

[17] De Bernardo M, Capasso L, Tortori A, Lanza M, Caliendo L and Rosa N 2014 Trans epithelial corneal collagen crosslinking for progressive keratoconus: 6 months follow up *Cont. Lens Anterior Eye* **37** 438–41

[18] Piñero D P and Alcón N 2014 *In vivo* characterization of corneal biomechanics *J. Cataract Refract. Surg.* **40** 870–87

[19] Rosa N, De Bernardo M, Iaccarino S and Lanza M 2015 Corneal biomechanical changes after myopic photorefractive keratectomy *Semin. Ophthalmol.* **30** 328–34

[20] Lanza M, Iaccarino S, Cennamo M, Irregolare C, Romano V and Carnevale U A 2015 Comparison between Corvis and other tonometers in healthy eyes *Cont. Lens Anterior Eye* **38** 94–8

[21] Vellara H R and Patel D V 2015 Biomechanical properties of the keratoconic cornea: a review *Clin. Exp. Optom.* **98** 31–8

[22] Rosa N, Lanza M, De Bernardo M, Signoriello G and Chiodini P 2015 Relationship between corneal hysteresis and corneal resistance factor with other ocular parameters *Semin. Ophthalmol.* **30** 335–9

[23] Lanza M, Iaccarino S and Bifani M 2016 *In vivo* human corneal deformation analysis with a Scheimpflug camera, a critical review *J. Biophotonics* **9** 464–77

[24] Hon Y and Lam A K 2013 Corneal deformation measurement using Scheimpflug noncontact tonometry *Optom. Vis. Sci.* **90** e1–8

[25] Hong J, Xu J, Wei A, Deng S X, Cui X, Yu X and Sun X 2013 A new tonometer—the Corvis ST tonometer: clinical comparison with noncontact and Goldmann applanation tonometers *Invest. Ophthalmol. Vis. Sci.* **54** 659–65

[26] Oculus Corvis® ST Pocket book (G/72100/0513/en) Wetzlar Germany 20013

[27] Han Z, Tao C, Zhou D, Sun Y, Zhou C, Ren Q and Roberts C J 2014 Air puff induced corneal vibrations: theoretical simulations and clinical observations *J. Refract. Surg.* **30** 208–13

[28] Huseynova T, Waring G O 4th, Roberts C, Krueger R R and Tomita M 2014 Corneal biomechanics as a function of intraocular pressure and pachymetry by dynamic infrared signal and Scheimpflug imaging analysis in normal eyes *Am. J. Ophthalmol.* **157** 885–93

[29] Tejwani S, Shetty R, Kurien M, Dinakaran S, Ghosh A and Sinha Roy A 2014 Biomechanics of the cornea evaluated by spectral analysis of waveforms from ocular response analyzer and Corvis-ST *PLoS One* **9** e97591

[30] Rabinowitz Y S 1998 Keratoconus *Surv. Ophthalmol.* **42** 297–319

[31] Keane M, Coster D, Ziaei M and Williams K 2014 Deep anterior lamellar keratoplasty versus penetrating keratoplasty for treating keratoconus *Cochrane Database Syst Rev.* CD009700

[32] Sorkin N and Varssano D 2014 Corneal collagen crosslinking: a systematic review *Ophthalmologica* **232** 10–27

[33] Khandelwal S S and Randleman J B 2015 Current and future applications of corneal cross-linking *Curr. Opin. Ophthalmol.* **26** 206–13

[34] Reynolds A, Moore J E, Naroo S A, Moore C B and Shah S 2010 Excimer laser surface ablation - a review *Clin. Exp. Ophthalmol.* **38** 168–82

[35] Kymionis G D, Kankariya V P, Plaka A D and Reinstein D Z 2012 Femtosecond laser technology in corneal refractive surgery: a review *J. Refract. Surg.* **28** 912–20

[36] Barsam A and Allan B D 2014 Excimer laser refractive surgery versus phakic intraocular lenses for the correction of moderate to high myopia *Cochrane Database Syst Rev.* CD007679

[37] O'Brart D P 2014 Excimer laser surface ablation: a review of recent literature *Clin. Exp. Optom.* **97** 12–7

[38] Rosa N, Capasso L, Lanza M and Romano A 2005 Axial eye length evaluation before and after myopic photorefractive keratectomy *J. Refract. Surg.* **21** 281–7

[39] Rosa N, De Bernardo M, Borrelli M, Filosa M L, Minutillo E and Lanza M 2011 Reliability of the IOLMaster in measuring corneal power changes after hyperopic photo-refractive keratectomy *J. Refract. Surg.* **27** 293–8

[40] Rosa N, Lanza M, Capasso L, Lucci M, Polito B and Romano A 2006 Anterior chamber depth measurement before and after photorefractive keratectomy: comparison between IOL master and Orbscan II *Ophthalmology* **113** 962–9

[41] Bromley J G and Randleman J B 2010 Treatment strategies for corneal ectasia *Curr. Opin. Ophthalmol.* **21** 255–8

[42] Hodge C, Chan C, Bali S J and Sutton G 2013 A review of corneal melting following kerato-refractive surgery *Clin. Exp. Optom.* **96** 14–9

[43] Parikh N B 2014 Management of residual refractive error after laser *in situ* keratomileusis and photorefractive keratectomy *Curr. Opin. Ophthalmol.* **25** 275–80

[44] Shah D N and Melki S 2014 Complications of femtosecond-assisted laser *in situ* keratomileusis flaps *Semin. Ophthalmol.* **29** 363–75

[45] Yildirim A, Cakir H, Kara N, Uslu H, Gurler B, Ozgurhan E B and Colak H N 2014 Corneal collagen crosslinking for ectasia after laser *in situ* keratomileusis: long-term results *J. Cataract Refract. Surg.* **40** 1591–6

[46] O'Brart D P, Shalchi Z, McDonald R J, Patel P, Archer T J and Marshall J 2014 Twenty-year follow-up of a randomized prospective clinical trial of excimer laser photorefractive keratectomy *Am. J. Ophthalmol.* **158** 651–63

[47] Chan T C, Liu D, Yu M and Jhanji V 2015 Longitudinal evaluation of posterior corneal elevation after laser refractive surgery using swept-source optical coherence tomography *Ophthalmology* **122** 687–92

[48] Yu A Y, Duan S F, Zhao Y E, Li X-Y, Lu F, Wang J and Wang Q-M 2011 Correlation between corneal biomechanical properties, applanation tonometry and direct intracameral tonometry *Br. J. Ophthalmol.* **96** 640–4

[49] Reinstein D Z, Gobbe M and Archer T J 2011 Ocular biomechanics: measurement parameters and terminology *J. Refract. Surg.* **27** 396–7

[50] Shields M B 1980 The non-contact tonometer. Its value and limitations *Surv. Ophthalmol.* **24** 211–9

[51] Spoerl E, Terai N, Scholz F, Raiskup F and Pillunat L E 2011 Detection of biomechanical changes after corneal cross-linking using ocular response analyzer software *J. Refract. Surg.* **27** 452–7

[52] Kynigopoulos M, Schlote T, Kotecha A, Tzamalis A, Pajic B and Haefliger I 2008 Repeatability of intraocular pressure and corneal biomechanical properties measurements by the ocular response analyser *Klin. Monbl. Augenheilkd.* **225** 357–60

[53] Moreno-Montañés J, Maldonado M J, García N, Mendiluce L, García-Gómez P J and Seguí-Gómez M 2008 Reproducibility and clinical relevance of the ocular response analyzer in nonoperated eyes: corneal biomechanical and tonometric implications *Invest. Ophthalmol. Vis. Sci.* **49** 968–74

[54] Bak-Nielsen S, Pedersen I B, Ivarsen A and Hjortdal J 2014 Dynamic Scheimpflug-based assessment of keratoconus and the effects of corneal cross-linking *J. Refract. Surg.* **30** 408–14

[55] Bak-Nielsen S, Pedersen I B, Ivarsen A and Hjortdal J 2015 Repeatability, reproducibility, and age dependency of dynamic Scheimpflug-based pneumotonometer and its correlation with a dynamic bidirectional pneumotonometry device *Cornea* **34** 71–7

[56] Pedersen I B, Bak-Nielsen S, Vestergaard A H, Ivarsen A and Hjortdal J 2014 Corneal biomechanical properties after LASIK, ReLEx flex, and ReLEx smile by Scheimpflug-based dynamic tonometry *Graefes Arch. Clin. Exp. Ophthalmol.* **252** 1329–35

[57] Chen X, Stojanovic A, Hua Y, Eidet J R, Hu D, Wang J and Utheim T P 2014 Reliability of corneal dynamic scheimpflug analyser measurements in virgin and post-PRK eyes *PLoS One* **9** e109577

[58] Ye C, Yu M, Lai G and Jhanji V 2015 Variability of corneal deformation response in normal and keratoconic eyes *Optom. Vis. Sci.* **92** e149–53

[59] Kamiya K, Hagishima M, Fujimura F and Shimizu K 2008 Factors affecting corneal hysteresis in normal eyes *Graefes Arch. Clin. Exp. Ophthalmol.* **246** 1491–4

[60] Touboul D, Roberts C, Kérautret J, Garra C, Maurice-Tison S, Saubusse E and Colin J 2008 Correlations between corneal hysteresis, intraocular pressure, and corneal central pachymetry *J. Cataract Refract. Surg.* **34** 616–22

[61] Lu F, Xu S, Qu J, Shena M, Wang X, Fang H and Wang J 2007 Central corneal thickness and corneal hysteresis during corneal swelling induced by contact lens wear with eye closure *Am. J. Ophthalmol.* **143** 616–22

[62] Shah S, Laiquzzaman M, Cunliffe I and Mantry S 2006 The use of the Reichert ocular response analyser to establish the relationship between ocular hysteresis, corneal resistance factor and central corneal thickness in normal eyes *Cont. Lens Anterior Eye* **29** 257–62

[63] Carbonaro F, Andrew T, Mackey D A, Spector T D and Hammond C J 2008 The heritability of corneal hysteresis and ocular pulse amplitude: a twin study *Ophthalmology* **115** 1545–9

[64] del Buey M A, Cristóbal J A, Ascaso F J, Lavilla L and Lanchares E 2009 Biomechanical properties of the cornea in Fuchs' corneal dystrophy *Invest. Ophthalmol. Vis. Sci.* **50** 3199–202

[65] Mangouritsas G, Morphis G, Mourtzoukos S and Feretis E 2009 Association between corneal hysteresis and central corneal thickness in glaucomatous and non-glaucomatous eyes *Acta Ophthalmol.* **87** 901–5

[66] Abitbol O, Bouden J, Doan S, Hoang-Xuan T and Gatinel D 2010 Corneal hysteresis measured with the ocular response analyzer in normal and glaucomatous eyes *Acta Ophthalmol.* **88** 116–9

[67] Alhamad T A and Meek K M 2011 Comparison of factors that influence the measurement of corneal hysteresis *in vivo* and *in vitro Acta Ophthalmol.* **89** e443–50

[68] Bayoumi N H, Bessa A S and El Massry A A 2010 Ocular response analyzer and Goldmann applanation tonometry: a comparative study of findings *J. Glaucoma* **19** 627–31

[69] Mansouri K, Leite M T, Weinreb R N, Tafreshi A, Zangwill L M and Medeiros F A 2012 Association between corneal biomechanical properties and glaucoma severity *Am. J. Ophthalmol.* **153** 419–27

[70] Galletti J G, Pfortner T and Bonthoux F F 2012 Improved keratoconus detection by ocular response analyzer testing after consideration of corneal thickness as a confounding factor *J. Refract. Surg.* **28** 1–7

[71] Hirneiss C, Neubauer A S, Yu A, Kampik A and Kernt M 2011 Corneal biomechanics measured with the ocular response analyser in patients with unilateral open-angle glaucoma *Acta Ophthalmol.* **89** e189–92

[72] Foster P J, Broadway D C, Garway-Heath D F, Yip J L Y, Luben R, Hayat S, Dalzell N, Wareham N J and Khaw K-T 2011 Intraocular pressure and corneal biomechanics in an adult British population: the EPIC-Norfolk eye study *Invest. Ophthalmol. Vis. Sci.* **52** 8179–85

[73] McMonnies C W 2012 Assessing corneal hysteresis using the ocular response analyzer *Optom. Vis. Sci.* **89** E343–9.

[74] Lim L, Gazzard G, Chan Y H, Fong A, Kotecha A, Sim E-L, Tan D, Tong L and Saw S-M 2008 Cornea biomechanical characteristics and their correlates with refractive error in Singaporean children *Invest. Ophthalmol. Vis. Sci.* **49** 3852–7

[75] Shen M, Fan F, Xue A, Wang J, Zhou X and Lu F 2008 Biomechanical properties of the cornea in high myopia *Vis. Res.* **48** 2167–71

[76] Plakitsi A, O'Donnell C, Miranda M A, Charman W N and Radhakrishnan H 2011 Corneal biomechanical properties measured with the ocular response analyser in a myopic population *Ophthalmic Physiol. Opt.* **31** 404–12

[77] Valbon B F, Ambrósio R Jr, Fontes B M, Luz A, Roberts C J and Alves M R 2014 Ocular biomechanical metrics by CorVis ST in healthy Brazilian patients *J. Refract. Surg.* **30** 468–73

[78] Lanza M, Cennamo M, Iaccarino S, Romano V, Bifani M, Irregolare C and Lanza A 2015 Evaluation of corneal deformation analyzed with a Scheimpflug based device *Cont. Lens Anterior Eye* **38** 89–93

[79] Mikielewicz M, Kotliar K, Barraquer R I and Michael R 2011 Air-pulse corneal applanation signal curve parameters for the characterization of keratoconus *Br. J. Ophthalmol.* **95** 793–8

[80] Shah S, Laiquzzaman M, Bhojwani R, Mantry S and Cunliffe I 2007 Assessment of the biomechanical properties of the cornea with the ocular response analyzer in normal and keratoconic eyes *Invest. Ophthalmol. Vis. Sci.* **48** 3026–31

[81] Fontes B M, Ambrósio R Jr, Jardim D, Velarde G C and Nosé W 2010 Corneal biomechanical metrics and anterior segment parameters in mild keratoconus *Ophthalmology* **117** 673–9

[82] Saad A, Lteif Y, Azan E and Gatinel D 2010 Biomechanical properties of keratoconus suspect eyes *Invest. Ophthalmol. Vis. Sci.* **51** 2912–6

[83] Fontes B M, Ambrósio R Jr, Velarde G C and Nosé W 2011 Ocular response analyzer measurements in keratoconus with normal central corneal thickness compared with matched normal control eyes *J. Refract. Surg.* **27** 209–15

[84] Hager A, Wegscheider K and Wiegand W 2009 Changes of extracellular matrix of the cornea in diabetes mellitus *Graefes Arch. Clin. Exp. Ophthalmol.* **247** 1369–74

[85] Castro D P, Prata T S, Lima V C, Biteli L G, de Moraes C G and Paranhos A Jr 2010 Corneal viscoelasticity differences between diabetic and nondiabetic glaucomatous patients *J. Glaucoma* **19** 341–3

[86] Kotecha A, Oddone F, Sinapis C, Elsheikh A, Sinapis D, Sinapis A and Garway-Heath D F 2010 Corneal biomechanical characteristics in patients with diabetes mellitus *J. Cataract Refract. Surg.* **36** 1822–8

[87] Sahin A, Bayer A, Ozge G and Mumcuoglu T 2009 Corneal biomechanical changes in diabetes mellitus and their influence on intraocular pressure measurements *Invest. Ophthalmol. Vis. Sci.* **50** 4597–604

[88] Pokharna H K and Pottenger L A 1997 Nonenzymatic glycation of cartilage proteoglycans: an *in vivo* and *in vitro* study *Glycoconj. J.* **14** 917–23

[89] Scheler A, Spoerl E and Boehm A 2012 Effect of diabetes mellitus on corneal biomechanics and measurement of intraocular pressure *Acta Ophthalmol.* **90** e447–51

[90] Hager A, Loge K, Füllhas M O, Schroeder B, Grossherr M and Wiegand W 2007 Changes in corneal hysteresis after clear corneal cataract surgery *Am. J. Ophthalmol.* **144** 341–6

[91] Kucumen R B, Yenerel N M, Gorgun E, Kulacoglu D N, Oncel B, Kohen M C and Alimgil M L 2008 Corneal biomechanical properties and intraocular pressure changes after phacoemulsification and intraocular lens implantation *J. Cataract Refract. Surg.* **34** 2096–8

[92] Mangouritsas G, Morphis G, Mourtzoukos S and Feretis E 2009 Association between corneal hysteresis and central corneal thickness in glaucomatous and non-glaucomatous eyes *Acta Ophthalmol.* **87** 901–5

[93] Ang G S, Bochmann F, Townend J and Azuara-Blanco A 2008 Corneal biomechanical properties in primary open angle glaucoma and normal tension glaucoma *J. Glaucoma* **17** 259–62

[94] Congdon N G, Broman A T, Bandeen-Roche K, Grover D and Quigley H A 2006 Central corneal thickness and corneal hysteresis associated with glaucoma damage *Am. J. Ophthalmol.* **141** 868–75

[95] Anand A, De Moraes C G, Teng C C, Tello C, Liebmann J M and Ritch R 2010 Corneal hysteresis and visual field asymmetry in open angle glaucoma *Invest. Ophthalmol. Vis. Sci.* **51** 6514–8

[96] Iordanidou V, Hamard P, Gendron G, Labbé A, Raphael M and Baudouin C 2010 Modifications in corneal biomechanics and intraocular pressure after deep sclerectomy *J. Glaucoma* **19** 252–6

[97] Sun L, Shen M, Wang J, Fang A, Xu A, Fang H and Lua F 2009 Recovery of corneal hysteresis after reduction of intraocular pressure in chronic primary angle-closure glaucoma *Am. J. Ophthalmol.* **147** 1061–6

[98] Tian L, Huang Y F, Wang L Q, Bai H, Wang Q, Jiang J J, Wu Y and Gao M 2014 Corneal biomechanical assessment using corneal visualization scheimpflug technology in keratoconic and normal eyes *J. Ophthalmol.* **2014** 147516

[99] Ali N Q, Patel D V and McGhee C N 2014 Biomechanical responses of healthy and keratoconic corneas measured using a noncontact scheimpflug-based tonometer *Invest. Ophthalmol. Vis. Sci.* **55** 3651–9

[100] Lanza M, Cennamo M, Iaccarino S, Irregolare C, Rechichi M, Bifani M and Gironi Carnevale U A 2014 Evaluation of corneal deformation analyzed with Scheimpflug based device in healthy eyes and diseased ones *Biomed. Res. Int.* **2014** 748671

[101] Pérez-Rico C, Gutiérrez-Ortíz C, González-Mesa A, Zandueta A M, Moreno-Salgueiro A and Germain F 2015 Effect of diabetes mellitus on Corvis ST measurement process *Acta Ophthalmol.* **93** e193–8

[102] Lee R, Chang R T, Wong I Y, Lai J S, Lee J W and Singh K 2016 Novel parameter of corneal biomechanics that differentiate normals from glaucoma *J. Glaucoma* **25** e603–9

[103] Chen M C, Lee N, Bourla N and Hamilton D R 2008 Corneal biomechanical measurements before and after laser *in situ* keratomileusis *J. Cataract Refract. Surg.* **34** 1886–91

[104] de Medeiros F W, Sinha-Roy A, Alves M R, Wilson S E and Dupps W J Jr 2010 Differences in the early biomechanical effects of hyperopic and myopic laser *in situ* keratomileusis *J. Cataract Refract. Surg.* **36** 947–53

[105] Shah S and Laiquzzaman M 2009 Comparison of corneal biomechanics in pre and post-refractive surgery and keratoconic eyes by ocular response analyser *Cont. Lens Anterior Eye* **32** 129–32

[106] Shah S, Laiquzzaman M, Yeung I, Pan X and Roberts C 2009 The use of the ocular response analyser to determine corneal hysteresis in eyes before and after excimer laser refractive surgery *Cont. Lens Anterior Eye* **32** 123–8

[107] Kamiya K, Shimizu K and Ohmoto F 2009 Time course of corneal biomechanical parameters after laser *in situ* keratomileusis *Ophthalmic Res.* **42** 167–71

[108] Dauwe C, Touboul D, Roberts C J, Mahmoud A M, Kérautret J, Fournier P, Malecaze F and Colin J 2009 Biomechanical and morphological corneal response to placement of intrastromal corneal ring segments for keratoconus *J. Cataract Refract. Surg.* **35** 1761–7

[109] Laiquzzaman M, Tambe K and Shah S 2010 Comparison of biomechanical parameters in penetrating keratoplasty and normal eyes using the ocular response analyser *Clin. Exp. Ophthalmol.* **38** 758–63

[110] Jafarinasab M R, Feizi S, Javadi M A and Hashemloo A 2011 Graft biomechanical properties after penetrating keratoplasty versus deep anterior lamellar keratoplasty *Curr. Eye Res.* **36** 417–21

[111] Tomita M, Mita M and Huseynova T 2014 Accelerated versus conventional corneal collagen crosslinking *J. Cataract Refract. Surg.* **40** 1013–20

[112] Maeda N, Ueki R, Fuchihata M, Fujimoto H, Koh S and Nishida K 2014 Corneal biomechanical properties in 3 corneal transplantation techniques with a dynamic Scheimpflug analyzer *Jpn. J. Ophthalmol.* **58** 483–9

[113] Shen Y, Zhao J, Yao P, Miao H, Niu L, Wang X and Zhou X 2014 Changes in corneal deformation parameters after lenticule creation and extraction during small incision lenticule extraction (SMILE) procedure *PLoS One* **9** e103893

[114] Mastropasqua L, Calienno R, Lanzini M, Colasante M, Mastropasqua A, Mattei P A and Nubile M 2014 Evaluation of corneal biomechanical properties modification after small incision lenticule extraction using Scheimpflug-based noncontact tonometer *Biomed. Res. Int.* **2014** 290619

[115] Hassan Z, Modis L Jr, Szalai E, Berta A and Nemeth G 2014 Examination of ocular biomechanics with a new Scheimpflug technology after corneal refractive surgery *Cont. Lens Anterior Eye* **37** 337–41

[116] Koprowski R, Lyssek-Boron A, Nowinska A, Wylegala E, Kasprzak H and Wrobel Z 2014 Selected parameters of the corneal deformation in the Corvis tonometer *Biomed. Eng. Online* **13** 55

[117] Koprowski R 2014 Automatic method of analysis and measurement of additional parameters of corneal deformation in the Corvis tonometer *Biomed. Eng. Online* **13** 150

[118] Koprowski R, Wilczyński S, Nowinska A, Lyssek-Boron A, Teper S, Wylegala E and Wróbel Z 2015 Quantitative assessment of responses of the eyeball based on data from the Corvis tonometer *Comput. Biol. Med.* **58** 91–100

[119] Koprowski R, Ambrósio R Jr and Reisdorf S 2015 Scheimpflug camera in the quantitative assessment of reproducibility of high-speed corneal deformation during intraocular pressure measurement *J. Biophotonics* **8** 968–78

[120] Koprowski R 2015 Open source software for the analysis of corneal deformation parameters on the images from the Corvis tonometer *Biomed. Eng. Online* **14** 31

[121] Ji C, Yu J, Li T, Tian L, Huang Y, Wang Y and Zheng Y 2015 Dynamic curvature topography for evaluating the anterior corneal surface change with Corvis ST *Biomed. Eng. Online* **14** 53

[122] Tian L, Ko M W, Wang L K, Zhang J Y, Li T J, Huang Y F and Zheng Y P 2014 Assessment of ocular biomechanics using dynamic ultra high-speed Scheimpflug imaging in keratoconic and normal eyes *J. Refract. Surg.* **30** 785–91

[123] Lanza M, Borrelli M, De Bernardo M, Filosa M L and Rosa N 2008 Corneal parameters and difference between goldmann applanation tonometry and dynamic contour tonometry in normal eyes *J. Glaucoma* **17** 460–4

[124] Broman A T, Congdon N G, Bandeen-Roche K and Quigley H A 2007 Influence of corneal structure, corneal responsiveness, and other ocular parameters on tonometric measurement of intraocular pressure *J. Glaucoma* **16** 581–8

[125] Rosa N, Lanza M, Borrelli M, De Bernardo M, Palladino A, Di Gregorio M G, Pascotto F and Politano L 2011 Low intraocular pressure resulting from ciliary body detachment in patients with myotonic dystrophy *Ophthalmology* **118** 260–4

[126] Lesk M R, Hafez A S and Descovich D 2006 Relationship between central corneal thickness and changes of optic nerve head topography and blood flow after intraocular pressure reduction in open-angle glaucoma and ocular hypertension *Arch. Ophthalmol.* **124** 1568–72

[127] Rosa N, Lanza M, de Bernardo M, Cecio M R, Passamano L and Politano L 2013 Intraocular pressure in patients with muscular dystrophies *Ophthalmology* **120** 1306–7

[128] Rosa N, De Bernardo M, Borrelli M, Filosa M L and Lanza M 2011 Effect of oxybuprocaine eye drops on corneal volume and thickness measurements *Optom. Vis. Sci.* **88** 640–4

Chapter 3

Clinical applications of the Corvis ST for glaucoma

**Marcelo Macedo, Marcelo Hatanaka, Hamed Momemi-Moghaddam
Remo Susanna Jr and Renato Ambrósio Jr**

3.1 Introduction

Glaucoma is the leading cause of irreversible blindness [1]. It affects more than 70 million people around the world, and it is estimated that it will reach 111.8 million in 2040 [2]. Elevated intraocular pressure (IOP) is the main risk factor for the development and progression of glaucoma. Therefore, the correct measurement of IOP is an essential part of diagnosis and for guiding the treatment and control of the disease [1, 2].

For many decades, the gold standard for measuring IOP has been Goldmann's applanation tonometry (GAT) [3]. However, there has been an increasing interest in the development of new modalities for IOP reading. One of the reasons for this greater attention towards tonometry started with the confirmation that the central cornea thickness (CCT), along with corneal biomechanical properties, do significantly affect measurements attained with GAT [4, 5]. In addition, the ocular hypertension treatment study (OHTS) was the first to demonstrate prospectively that CCT is also an independent risk factor for patients with ocular hypertension developing glaucomatous neuropathy [6]. The model that combines both OHTS and EGPS (European Glaucoma Prevention Study Group) data, calculated that for every 40 μm of CCT decrease, there is a two times greater risk of developing a glaucomatous lesion in five years [7]. It was also verified that the precision of applanation tonometry was influenced (besides by the thickness of the cornea) by the rigidity, curvature and age of the cornea [8]. The influence of CCT in measuring IOP was assessed in numerous studies [9–11]. However, there is still a lack of an adequate method to surpass GAT limitations and measure the real IOP.

In simple terms, Goldmann's tonometry underestimates the IOP in eyes with thinner corneas and overestimates IOP in eyes with thicker corneas. As a consequence, based on IOP measurements in eyes with thinner corneas, patients with primary open-glaucoma (POAG) may be erroneously classified as normal subjects. On the other hand, normotensive eyes with thicker corneas may classified as hypertensive. This is a particularly important matter, given the increasing number of refractive surgery patients with thinner corneas [8, 10, 11].

Many studies have evaluated the effects of CCT on IOP measurements performed with GAT. A wide array of estimates between 0.7 and 7.1 mmHg for each 100 μm of corneal alteration has been described [4, 12–16]. Interestingly, the relationship between IOP and corneal shape, thickness and biomechanical properties was mathematically demonstrated by Liu and Roberts [17]; it being much more complex than a simple linear nature. This research demonstrated that variations in the biomechanics of the cornea have the potential to produce an error in measuring the IOP of 17 mmHg at the extremes of the Young Corneas modules, compared with errors in measuring the magnitude of 2–3 mmHg in the normal range values of CCT and curvature. The relationship between IOP and corneal properties is even more complex when taking variations of corneal elasticity into account; the thickness effects being more pronounced in eyes with more rigid corneas in comparison to more flexible corneas. Furthermore, the authors concluded that the variations of biomechanical properties of the cornea between individuals have a much bigger impact on IOP than the thickness or curvature of the cornea [17]. In addition, Kwon and colleagues [18] showed that the biomechanical properties of the cornea characterized by the rigidity parameter were as important as the CCT in influencing the IOP measured by the Goldmann tonometer. Thus, studies have increasingly shown that the effects of the cornea on IOP measurements with the Goldmann tonometer isn't limited to the thickness of the cornea [12, 15, 17, 18].

The cornea has viscoelastic properties that can be evaluated through biomechanical measurements [19]. Abnormal corneal biomechanics is correlated with various ocular and systemic conditions [20] such as keratoconus, Fuchs dystrophy, diabetes mellitus, and can also be used to influence surgical results [17], especially refractive surgery. Moreover, corneal biomechanics also affects measuring the IOP, leading to imprecise tonometry readings [21] and may be considered an independent risk factor for the development of glaucoma [22–25].

This led to the development of new tonometers that try to assess the IOP in a way that is independent of these properties of the cornea, such as the ocular response analyzer (ORA, Reichert Inc, Depew NY) [26], and the Corvis ST (Corvis ST, Oculus Optikgeräte GmbH, Wetzlar, Germany) [27]. This chapter highlights the current applications of the Corvis ST for clinical evaluation in glaucoma.

3.2 Basic description of the Corvis ST

The Corvis ST is a non-contact tonometer [28] that monitors the cornea with an ultra-high-speed Scheimpflug camera that allows one to record the corneal deformation response for *in vivo* characterization of biomechanical parameters of the cornea along with evaluating IOP and CCT [29]. The Corvis ST utilizes a constant collimated air jet to deform the cornea. The ultra-high-velocity camera uses

Scheimpflug geometry to capture images of the horizontal meridian in more than 4300 frames per second, resulting in 140 images during a 31 ms air jet [27]. It was designed using numerical simulations and finite elements applied in human eye models, with different corneal shapes (3D tomography), including thickness, age and IOP value results [12, 30–33].

Along with the corneal response dynamic parameters, the Corvis ST provides a standard IOP, pachymetry and a new and validated estimation of biomechanical-corrected intraocular pressure (bIOP) intended to be an IOP measurement, which is not influenced by corneal thickness and/or rigidity parameters (figure 3.1) [34].

3.3 bIOP: concept and first clinical results

To analyze the parameters of the Corvis ST, a set of data was divided into groups of different bIOP, age, and CCT. A comparative analysis of the pachymetry subgroups did not present a significant difference in bIOP and age, but was significantly different for the uncorrected IOP. This result demonstrated that the bIOP correction algorithm is capable of compensating for these important confusing factors and confirms the pre-clinical validity of the formula created for this device [34].

The analyses of the relationship between the dynamic parameters of corneal response and the CCT demonstrated that the biggest radial concavity, inverse radial concavity, amplitude of deformation ratio and amplitude of deflection ratio were highly correlated with the CCT and were not significantly influenced by the IOP. This means that these are good parameters to assess correctly the corneal

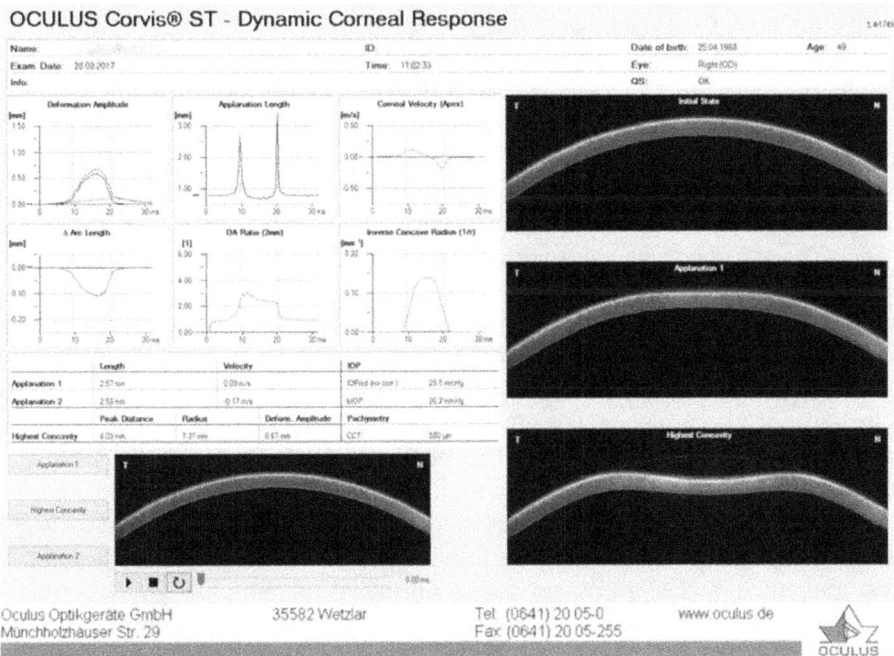

Figure 3.1. Corneal parameters as presented by the Corvis ST.

biomechanics *in vivo*, due to its relative independence from the IOP. Another important finding is the confirmation that many parameters used in other prior studies, such as the amplitude of deformation, are strongly correlated with the IOP [35, 36].

Other results of the study comparing age groups, showing difference in the bIOP, indicated slightly lower values the older the age group. Total ocular movement, deformation amplitude ratio and inverse concave radius were the three most strongly influenced parameters by age. The significant difference in terms of IOP must be considered with care, given the small changes in terms of bIOP and a literature that doesn't show any independent effect of age over IOP [37, 38].

These authors conclude that the correlation of the deformation amplitude ratio and the inverse concave radius with age, together with its pachymetry, probably indicates its capacity to quantify the corneas biomechanics [37, 38].

Tejwani *et al* [39] observed that the results obtained by the Corvis ST and ORA were statistically different. Bak-Nielsen *et al* [29] also showed that the parameters of the Corvis ST have a weak correlation with CH and CRF obtained by ORA measurements. Recently, a variety of studies indicate significant changes in the cornea's biomechanics, according to measurements made by the Corvis ST in various situations, such as keratoconus, dry eye, myopia, glaucoma, and refractive surgery [40–44].

To approach the effects of corneal rigidity in measuring the IOP, a recent study developed a correction equation based on numerical simulation with the Corvis ST. The equation was posteriorly validated using a set of clinical data involving 632 patients and demonstrated success in reducing significantly the association of Corvis ST with the IOP, CCT and age [34].

This has a deep impact in assessing the cornea's biomechanics *in vivo*, since the creation of an algorithm of the corrected IOP with a significant reduction in the CCT and influence of age, but which may influence rigidity, is the first step to assessing the cornea's biomechanics [34].

A recent study assessed corneal biomechanic parameters using the Corvis ST with different intraocular pressures in the same eye, submitted to the water drinking test (WDT). The WDT consists of drinking 800 mL of water at room temperature, in at least 5 min, after a minimum of 2 h free of water ingestion. The IOP measurements are made before drinking water and three other measurements, with an interval of 15 min between them, after water ingestion [45].

The test has been proposed as a tool to assess indirectly the facility of drainage of the reserve of the aqueous humor, besides being an indicator of risk for the progression of glaucoma [46, 47]. The peak of IOP, after water ingestion, has been described as an independent risk factor for the development or progression of the campimetric glaucomatous defect in patients with open angle glaucoma [47]. Eyes with asymmetric glaucomatous lesions and eyes with a greater campimetric defect presented higher pressuric peaks during WDT [46]. A good correlation and agreement with pressure peaks that normally occur during the day was also shown [48].

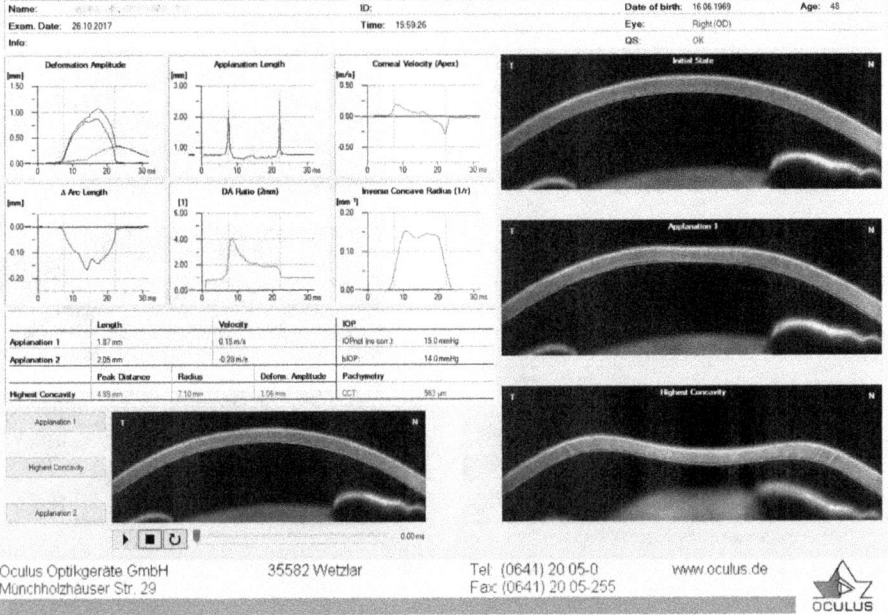

Figure 3.2. Initial intraocular pressure before water ingestion.

Figure 3.2 presents the initial IOP of a given patient. Then, IOP rise after water ingestion is presented in figure 3.3. Figure 3.4 presents all IOP measurements during each time point of the test. The deformation amplitude and velocity decline with the rise of IOP, despite the constant corneal thickness, demonstrates that it is not only corneal thickness that is an important variable in the assessment of the true IOP value [49]. Correia *et al* [50] have already reported that the composition of a given material may be more important than its thickness. Roberts *et al* have demonstrated that IOP has a bigger influence over corneal deformation amplitude than thickness and curvature [51]. Very importantly, we have found no difference in IOP levels measured with the Corvis ST before and after instilling fluorescein in 50 healthy volunteers [49].

When comparing bIOP measured by the Corvis ST (Oculus, Wetzlar, Germany) with different tonometers, such as the Goldmann digital applanation tonometer (GAT: Haag-Streit AG, Bern, Switzerland), Icare (Icare® TA01i, Icare Finland Oy, Finland), the manual ART (BioResonator ART manual), Topcon non-contact tonometer (Topcon NCT, Topcon Corporation, Tokyo, Japan) and even the ocular response analyzer (ORA: Reichert Ophthalmic Instruments, Buffalo, New York, USA), Momemi-Moghaddam *et al* [52] demonstrated that the use of conventional tonometers, especially in thin corneas and after corneal refractive surgery, might result in inaccurate IOP readings, which may impact future diagnosis such as in normotensive, glaucomatous or ocular hypertensive eyes.

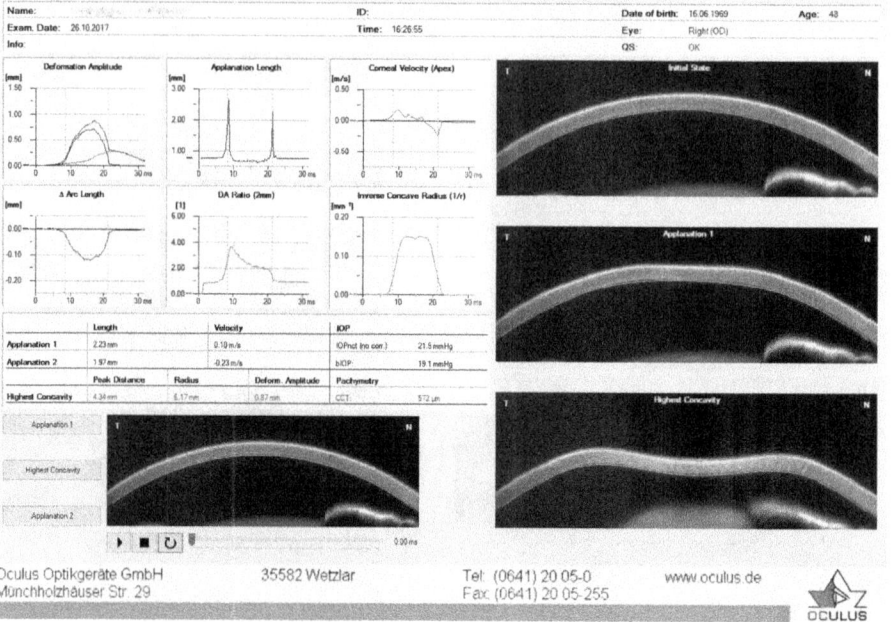

Figure 3.3. Intraocular pressure after water ingestion.

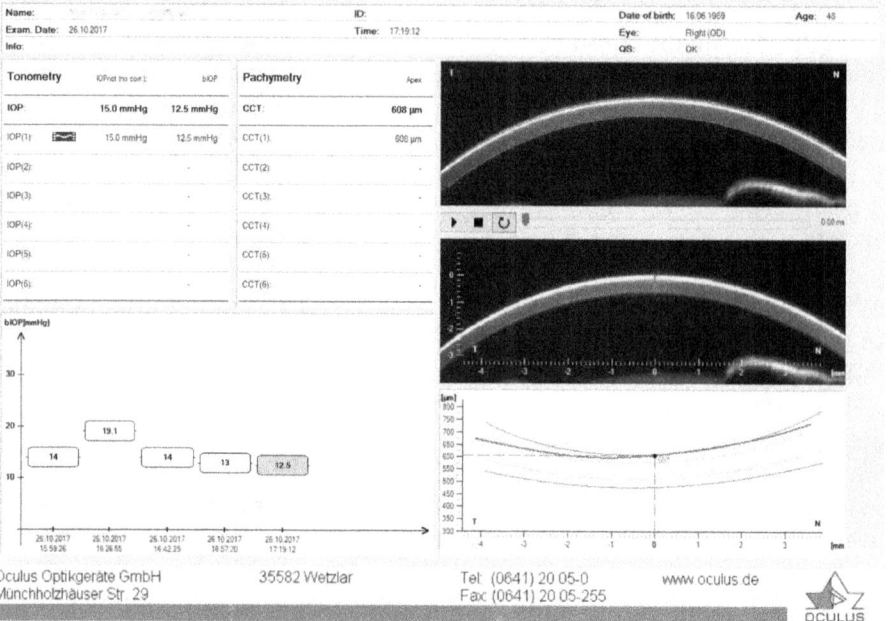

Figure 3.4. Intraocular pressure during the WDT. Peak IOP, 15 min after water ingestion, is detected.

Important conclusions, such as the ones made by the group of Vinciguerra *et al* [53], who analyzed the dynamic parameters of the corneal response along with literature concerning bIOP, CCT and age, confirm that IOP and CCT are important factors in assessing *in vivo* biomechanics with the influence of age. Higher concavity radius, inverse concave radius, deformation amplitude ratio and deflection amplitude ratio can be good parameters in assessing corneal biomechanics *in vivo* with relative independence to IOP and its correlation with CCT and age.

3.4 Conclusion

It would be interesting to obtain IOP values that are independent of corneal factors and know their resemblance to true IOP. Biomechanical factors of the cornea are being studied with the aim of making it possible not only to obtain the correct IOP but also to verify the importance of each factor in the development of glaucoma.

References

[1] Quigley H A and Broman A T 2006 The number of people with glaucoma worldwide in 2010 and 2020 *Br. J. Ophthalmol.* **90** 262–7

[2] Tham Y C, Li X, Wong T Y, Quigley H A, Aung T and Cheng C Y 2014 Global prevalence of glaucoma and projections of glaucoma burden through 2040: a systematic review and meta-analysis *Ophthalmology* **121** 2081–90

[3] Good T J, Kimura A E, Mandava N and Kahook M Y 2011 Sustained elevation of intraocular pressure after intravitreal injections of anti-VEGF agents *Br. J. Ophthalmol.* **95** 1111–4

[4] Ehlers N, Bramsen T and Sperling S 1975 Applanation tonometry and central corneal thickness *Acta Ophthalmol.* **53** 34–43

[5] Whitacre M M, Stein R A and Hassanein K 1993 The effect of corneal thickness on applanation tonometry *Am. J. Ophthalmol.* **115** 592–6

[6] Gordon M O, Beiser J A, Brandt J D, Heuer D K, Higginbotham E J and Johnson C A *et al* 2002 The ocular hypertension treatment study: baseline factors that predict the onset of primary open-angle glaucoma *Arch. Ophthalmol.* **120** 714–20 discussion 829–30

[7] Ocular Hypertension Treatment Study Group *et al* 2007 Validated prediction model for the development of primary open-angle glaucoma in individuals with ocular hypertension *Ophthalmology* **114** 10–9

[8] Elsheikh A, Gunvant P, Jones S W, Pye D and Garway-Heath D 2013 Correction factors for Goldmann tonometry *J. Glaucoma* **22** 156–63

[9] Joda A A, Shervin M M S, Kook D and Elsheikh A 2016 Development and validation of a correction equation for Corvis tonometry *Comput. Methods Biomech. Biomed. Eng.* **19** 943–53

[10] Kohlhaas M, Boehm A G, Spoerl E, Pursten A, Grein H J and Pillunat L E 2006 Effect of central corneal thickness, corneal curvature, and axial length on applanation tonometry *Arch. Ophthalmol.* **124** 471–6

[11] Spoerl E, Terai N and Pillunat L E 2012 Age-dependent correction factors for Goldmann tonometry *J. Glaucoma* **21** 276–7

[12] Elsheikh A, Alhasso D, Gunvant P and Garway-Heath D 2011 Multiparameter correction equation for Goldmann applanation tonometry *Optom. Vis. Sci.* **88** E102–12

[13] Foster P J, Baasanhu J, Alsbirk P H, Munkhbayar D, Uranchimeg D and Johnson G J 1998 Central corneal thickness and intraocular pressure in a Mongolian population *Ophthalmology* **105** 969–73

[14] Gunvant P, Baskaran M, Vijaya L, Joseph I S, Watkins R J and Nallapothula M *et al* 2004 Effect of corneal parameters on measurements using the pulsatile ocular blood flow tonograph and Goldmann applanation tonometer *Br. J. Ophthalmol.* **88** 518–22

[15] Kotecha A, White E T, Shewry J M and Garway-Heath D F 2005 The relative effects of corneal thickness and age on Goldmann applanation tonometry and dynamic contour tonometry *Br. J. Ophthalmol.* **89** 1572–5

[16] Wolfs R C, Klaver C C, Vingerling J R, Grobbee D E, Hofman A and de Jong P T 1997 Distribution of central corneal thickness and its association with intraocular pressure: the Rotterdam study *Am. J. Ophthalmol.* **123** 767–72

[17] Liu J and Roberts C J 2005 Influence of corneal biomechanical properties on intraocular pressure measurement: quantitative analysis *J. Cataract Refract. Surg.* **31** 146–55

[18] Kwon T H, Ghaboussi J, Pecknold D A and Hashash Y 2010 Role of corneal biomechanical properties in applanation tonometry measurements *J. Refract. Surg.* **26** 512–9

[19] Pinero D P and Alcon N 2014 *In vivo* characterization of corneal biomechanics *J. Cataract Refract. Surg.* **40** 870–87

[20] Pinero D P and Alcon N 2015 Corneal biomechanics: a review *Clin. Exp. Optom.* **98** 107–16

[21] Roberts C 2005 Biomechanical customization: the next generation of laser refractive surgery *J. Cataract Refract. Surg.* **31** 2–5

[22] Brown K E and Congdon N G 2006 Corneal structure and biomechanics: impact on the diagnosis and management of glaucoma *Curr. Opin. Ophthalmol.* **17** 338–43

[23] Congdon N G, Broman A T, Bandeen-Roche K, Grover D and Quigley H A 2006 Central corneal thickness and corneal hysteresis associated with glaucoma damage *Am. J. Ophthalmol.* **141** 868–75

[24] Kotecha A 2007 What biomechanical properties of the cornea are relevant for the clinician? *Surv. Ophthalmol.* **52** S109–14

[25] Tejwani S, Devi S, Dinakaran S, Shetty R, Meshram P, Francis M and Sinha Roy A 2016 Diagnostic efficacy of normalization of corneal deformation variables by the intraocular pressure in glaucomatous eyes *Invest. Ophthalmol. Vis. Sci.* **57** 1082–6

[26] Luce D A 2005 Determining *in vivo* biomechanical properties of the cornea with an ocular response analyzer *J. Cataract Refract. Surg.* **31** 156–62

[27] Ambrósio R Jr, Ramos I, Luz A, Faria F C, Steinmueller A, Krug M, Belin M W and Roberts C J 2013 Dynamic ultra high speed Scheimpflug imaging for assessing corneal biomechanical properties *Rev. Bras. Oftalmol.* **72** 99–102

[28] Mussi C E, Colombo P, Lo Russo C, Kasangian A, Cananzi F and Marrari A *et al* 2016 Sporadic desmoid tumors of the abdominal wall: the results of surgery *Tumori* **102** 582–7

[29] Bak-Nielsen S, Pedersen I B, Ivarsen A and Hjortdal J 2015 Repeatability, reproducibility, and age dependency of dynamic Scheimpflug-based pneumotonometer and its correlation with a dynamic bidirectional pneumotonometry device *Cornea* **34** 71–7

[30] Davey P G, Elsheikh A and Garway-Heath D F 2013 Clinical evaluation of multiparameter correction equations for Goldmann applanation tonometry *Eye* **27** 621–9

[31] Elsheikh A 2010 Finite element modeling of corneal biomechanical behavior *J. Refract. Surg.* **26** 289–300

[32] Elsheikh A, Geraghty B, Rama P, Campanelli M and Meek K M 2010 Characterization of age-related variation in corneal biomechanical properties *J. R. Soc. Interface* **7** 1475–85

[33] Elsheikh A, Wang D, Brown M, Rama P, Campanelli M and Pye D 2007 Assessment of corneal biomechanical properties and their variation with age *Curr. Eye Res.* **32** 11–9

[34] Joda A A, Shervin M M, Kook D and Elsheikh A 2016 Development and validation of a correction equation for Corvis tonometry *Comput. Methods Biomech. Biomed. Eng.* **19** 943–53

[35] Bao F, Deng M, Wang Q, Huang J, Yang J, Whitford C, Geraghty B, Yu A and Elsheikh A 2015 Evaluation of the relationship of corneal biomechanical metrics with physical intraocular pressure and central corneal thickness in *ex vivo* rabbit eye globes *Exp. Eye Res.* **137** 11–7

[36] Huseynova T, Waring G O, Roberts C, Krueger R R and Tomita M 2014 Corneal biomechanics as a function of intraocular pressure and pachymetry by dynamic infrared signal and Scheimpflug imaging analysis in normal eyes *Am. J. Ophthalmol.* **157** 885–93

[37] Nomura H, Ando F, Niino N, Shimokata H and Miyake Y 2002 The relationship between age and intraocular pressure in a Japanese population: the influence of central corneal thickness *Curr. Eye Res.* **24** 81–5

[38] Rochtchina E, Mitchell P and Wang J J 2002 Relationship between age and intraocular pressure: the Blue Mountains eye study *Clin. Exp. Ophthalmol.* **30** 173–5

[39] Tejwani S, Shetty R, Kurien M, Dinakaran S, Ghosh A and Sinha Roy A 2014 Biomechanics of the cornea evaluated by spectral analysis of waveforms from ocular response analyzer and Corvis-ST *PLoS One* **9** e97591

[40] Long Q, Wang J, Yang X, Jin Y, Ai F and Li Y 2015 Assessment of corneal biomechanical properties by CorVis ST in patients with dry eye and in healthy subjects *J. Ophthalmol.* **2015** 380624

[41] Pedersen I B, Bak-Nielsen S, Vestergaard A H, Ivarsen A and Hjortdal J 2014 Corneal biomechanical properties after LASIK, ReLEx flex, and ReLEx smile by Scheimpflug-based dynamic tonometry *Graefes Arch. Clin. Exp. Ophthalmol.* **252** 1329–35

[42] Steinberg J, Katz T, Lucke K, Frings A, Druchkiv V and Linke S J 2015 Screening for keratoconus with new dynamic biomechanical *in vivo* Scheimpflug analyses *Cornea* **34** 1404–12

[43] Wang J, Li Y, Jin Y, Yang X, Zhao C and Long Q 2015 Corneal biomechanical properties in myopic eyes measured by a dynamic Scheimpflug analyzer *J. Ophthalmol.* **2015** 161869

[44] Wang W, Du S and Zhang X 2015 Corneal deformation response in patients with primary open-angle glaucoma and in healthy subjects analyzed by Corvis ST *Invest. Ophthalmol. Vis. Sci.* **56** 5557–65

[45] Medeiros F A, Pinheiro A, Moura F C, Leal B C and Susanna R Jr 2002 Intraocular pressure fluctuations in medical versus surgically treated glaucomatous patients *J. Ocul. Pharmacol. Ther.* **18** 489–98

[46] Susanna R Jr, Hatanaka M, Vessani R M, Pinheiro A and Morita C 2006 Correlation of asymmetric glaucomatous visual field damage and water-drinking test response *Invest. Ophthalmol. Vis. Sci.* **47** 641–4

[47] Susanna R Jr, Vessani R M, Sakata L, Zacarias L C and Hatanaka M 2005 The relation between intraocular pressure peak in the water drinking test and visual field progression in glaucoma *Br. J. Ophthalmol.* **89** 1298–301

[48] De Moraes C G, Furlanetto R L, Reis A S, Vegini F, Cavalcanti N F and Susanna R Jr 2009 Agreement between stress intraocular pressure and long-term intraocular pressure measurements in primary open angle glaucoma *Clin. Exp. Ophthalmol.* **37** 270–4

[49] Macedo M *et al* 2018 unpublished data

[50] Correia F F, Ramos I, Roberts C J, Steinmueller A, Krug M and Ambrosio R Jr 2013 Impact of chamber pressure and material properties on the deformation response of corneal models measured by dynamic ultra-high-speed Scheimpflug imaging *Arq. Bras. Oftalmol.* **76** 278–81

[51] Roberts C J, Mahmoud A M, Ramos I, Caldas D, da Silva R S and Ambrósio R Jr 2011 Factors influencing corneal deformation and estimation of intraocular pressure *Invest. Ophthalmol. Vis. Sci.* **52** 4384

[52] Momemi-Moghaddam H *et al* 2018 unpublished data

[53] Vinciguerra R, Elsheikh A, Roberts C J, Kang D S Y, Lopes B T, Morenghi E, Azzolini C and Vinciguerra P 2016 Influence of pachymetry and intraocular pressure on dynamic corneal response parameters in healthy patients *J. Refract. Surg.* **32** 550–61

Chapter 4

Evaluation of the algorithms utilised to diagnose keratoconus for the Corvis ST

Hans R Vellara, Akilesh Gokul, Charles N J McGhee and Dipika V Patel

4.1 Introduction

Since the commercial release of the Corvis ST (CST) (Oculus, Wetzlar, Germany), several authors have proposed various additional algorithms for analysing the CST video output for the purpose of diagnosing keratoconus [1–8]. However, to our knowledge, there have been no investigations comparing these algorithms to each other in a single cohort of patients.

The aim of this study was to assess and compare the efficacy of different algorithms in diagnosing keratoconus.

4.2 Methods

Informed written consent was obtained from all patients and participants, and the study design adhered to the tenets of the Declaration of Helsinki. The study protocol was approved prospectively by the University of Auckland Human Ethics Committee (reference number 010853).

4.2.1 Patients with keratoconus

All patients were recruited from the Corneal Service at Greenlane Clinical Centre, Auckland District Health Board, New Zealand.

The Rabinowitz [9] definition of keratoconus was utilised by experienced clinicians to diagnose the disease on the basis of computerised tomography. These features included an increased area of corneal power surrounded by concentric areas of decreasing power; skewing of the steepest radial axes above and below the horizontal meridian; and superior–inferior power asymmetry [9]. The severity of keratoconus was classified using the steepest keratometry (K_{MAX}) values as outlined in the Collaborative Longitudinal Evaluation of Keratoconus (CLEK) study [10]: mild (< 45 D), moderate (45–52 D), and severe keratoconus (> 52 D). Tomographic characteristics were assessed using the Pentacam HR as described in [10]. Eyes with stromal scarring, or a history of corneal hydrops or corneal surgery were excluded from the study.

4.2.2 Participants

Participants were recruited from the same tertiary referral centre as noted above. Baseline data for the cohort of healthy participants used in this study has been described [2].

Within our master cohort, a subset of keratoconic eyes and healthy participants (with matching age, gender, ethnicity, intraocular pressures, and corneal thickness) were selected and analysed for this study as these factors are known to influence the biomechanics of the eye.

Exclusion criteria for both groups were a previous history of ocular or periocular trauma and/or surgery, or any ocular or systemic conditions that may affect the ocular biomechanics.

4.2.3 Examinations

The examinations were performed by experienced investigators (HRV, and AG). A medical history was obtained and slit lamp biomicroscopy was performed on all patients to determine their ocular status. A Pentacam HR tomographer was utilised to measure central corneal thickness and maximum simulated keratometry. The biomechanical response was measured with the CST.

4.2.4 Additional algorithms

Six algorithms were selected for comparison. These were chosen based on their previous utilisation for comparison—specifically between healthy and keratoconic corneas. Furthermore, only quantitative parameters were included. The rationale behind each algorithm is briefly described below. These have been ordered alphabetically by the author's last name. The detailed methodology for each algorithm can be found in the corresponding references following each subheading.

Koprowski and Ambrósio Jr (frequency of first harmonic, minimum, and maximum corneal deformation for frequency over 150 Hz) [4, 11].

The inherent structural stiffness of a material can be described from forced vibrations. A stiffer material would vibrate at a higher frequency with a lower amplitude when agitated, with the opposite observed for a material of lower stiffness. Following this logic, keratoconic corneal tissue may have a lower corneal deformation frequency as it has lower elasticity compared to healthy corneas [12, 13]. Koprowski and Ambrósio Jr described algorithms that derived corneal vibrations from the CST video.

Shetty et al ('an' root mean square and area under the deformation amplitude curve) [3].

The viscous response from corneal tissue may be governed by the proteoglycan matrix produced by the keratocytes [14]. As the keratocytes are abnormal in keratoconus, the response to a cyclic load could be more asymmetric compared to a healthy cornea [14, 15]. Shetty *et al* [3] postulated an increase in imperfections in the Fourier fit of data correlated to the increased asymmetric response to an air pulse in keratoconic tissue. Furthermore, dynamic Scheimpflug analysis of keratoconic tissue has revealed several changes in the CST response compared to healthy corneal tissue [16]. This includes time, velocity, and magnitude changes in deformation amplitude over time. The area under the deformation amplitude over the time curve encapsulates all these changes and thus may provide a higher diagnostic value. The two additionally derived parameters that provided the highest area under the receiver operating characteristic analysis curve (AUC) were included in this study.

Steinberg *et al* (applanation length level and deflection length level) [6].

The inbuilt CST software-derived parameters are based on the evaluation of a single frame. Applanation length level and deflection length level consider the corneal biomechanical response over a dynamic range (several frames) and thus may better describe the dynamic corneal biomechanical properties.

Tian *et al* (maximum corneal inward velocity) [1].

Tian *et al* demonstrated the highest AUC in the diagnosis between healthy and keratoconic corneas with a new parameter termed 'maximum corneal inward velocity'. This reflected a reduced number of corneal collagen fibrils, altered orientation, and fewer crosslinks found in keratoconic tissue, which lead to a more fragile and easily extensible tissue.

Vellara *et al* (maximum corneal deformation and corneal energy dissipation).

The derivation of both of these parameters has been described in detail [2]. Maximum corneal deformation contains pure corneal deformation. Therefore, it may better reflect the biomechanical differences in diseases (such as keratoconus) that affect just the cornea. Furthermore, viscoelastic materials undergo a loss of energy under cyclic deformation (hysteresis) [14]. As explained above, the proteoglycan matrix could be abnormal in keratoconus, which may lead to a difference in hysteresis compared to healthy corneas. Corneal energy dissipation describes this loss of corneal energy.

Vinciguerra *et al* (Ambrósio's relational thickness to the horizontal profile, and corneal biomechanical index) [7].

In comparison to the healthy cornea, the thickness profile of a keratoconic cornea increases at a faster rate from the corneal centre to the corneal periphery. Ambrósio's relational thickness to the horizontal profile provides an indication of this increase. However, a lower value can indicate either a thin cornea and/or a faster increase in thickness towards the periphery. The corneal biomechanical index was empirically derived using a logistic regression analysis with forward stepwise inclusion, which combined various indices from the CST (including Ambrósio's relational thickness to the horizontal profile) as predictors for the specific diagnosis of keratoconus.

Wang *et al* (transient stiffness coefficient and energy absorbed area) [5].

Classical tissue biomechanical characteristics, such as elasticity and viscosity, allow direct material comparisons between tissues. The transient stiffness coefficient closely replicates the derivation of elasticity. However, the energy absorbed area reflects corneal hysteresis, which is postulated to be different between keratoconic and healthy corneas.

4.2.5 Statistical analysis

Statistical analysis was performed using Prism software version 6.01 (GraphPad, La Jolla, California, USA), and R software version 3.3.1 (R Foundation, Vienna, Austria).

Descriptive statistical analysis was performed on baseline characteristics. Only one eye from each participant was included in the analysis. The normality of the data was tested using the D'Agostino and Pearson omnibus normality test. An unpaired student's t-test (parametric) and a Kolmogorov–Smirnov test (non-parametric) were performed to identify variables that were significantly different between groups. The statistical differences between ethnicities were assessed using the chi-squared test.

A receiver operating characteristic analysis was performed to investigate the efficacy of each algorithm in diagnosing keratoconic corneas. The particular algorithm is considered poor when the AUC is in the range 0.5–0.7, and useful in the range 0.7–0.9. An AUC greater than 0.9 was considered good to be excellent (1 = perfect) [17]. The AUCs were compared for parameters that were significantly different between the groups [18]. P-values of 0.05 or less were considered to be statistically significant.

4.3 Results

4.3.1 Patient and participant demographics and descriptors

The master study group consists of two hundred eyes of 200 patients with keratoconus and a healthy cohort of 152 eyes of 152 participants. Two subgroups were derived from this master-set of ($N = 352$ subjects), with a separate subset of keratoconic eyes ($n = 100$) and healthy participants ($n = 90$). These two subgroups were created with matching age, gender, ethnicities, intraocular pressures, and corneal thickness for the purposes of this study. Mean age ± standard deviation (SD), gender, and ethnicity proportions, intraocular pressures, and central corneal thicknesses for these patients are displayed in table 4.1. The severity distribution of keratoconic eyes was as follows: mild ($n = 30$), moderate ($n = 56$), and severe ($n = 14$).

In the subset populations analysed; the mean, SD, and P-value for every algorithm have been provided in table 4.2. Six out of twelve investigated parameters were significantly different between groups and most were higher in the keratoconic group.

4.3.2 Receiver operating characteristic analysis

The algorithms that were significantly different between healthy and keratoconic groups were incorporated into a receiver operating characteristic analysis. The results of this analysis have been provided in table 4.3. The 'corneal biomechanical index' provided the highest AUC (0.93 ± 0.03) followed by 'Ambrósio's relational thickness to the horizontal profile' (0.91 ± 0.04). However, the corneal biomechanical index had an AUC that was not statistically significantly different from Ambrósio's relational thickness to the horizontal profile ($P = 0.31$). A cut-off value of 0.53 for the corneal biomechanical index yielded 88% sensitivity and 87%

Table 4.1. The mean age ± standard deviation (SD), gender proportions, intraocular pressures, central corneal thicknesses, and ethnicities in a subset of keratoconic patients ($n = 100$) and healthy participants ($n = 90$).

Parameter	Keratoconus	Healthy	P-value
Mean age ± SD (years)	28.4 ± 10.9	30.2 ± 11.0	0.25
Gender	52.2% females, $n = 47$	58.0% females, $n = 58$	0.22
intraocular pressure (mmHg)	11.9 ± 2.0	12.4 ± 2.0	0.10
Central corneal thickness (μm)	518 ± 31	525 ± 25	0.38
Ethnicity			
European	32.0%, $n = 32$	37.8%, $n = 34$	0.13
East/Southeast Asian	20.0%, $n = 20$	22.2%, $n = 20$	0.76
Indian subcontinent	18.0%, $n = 18$	23.3%, $n = 21$	0.23
Pacific island nations	13.0%, $n = 13$	6.7%, $n = 6$	0.07
Maori	10.0%, $n = 10$	5.6%, $n = 5$	0.09
Middle Eastern	5.0%, $n = 5$	2.2%, $n = 2$	0.06
African descent	2.0%, $n = 2$	3.3%, $n = 3$	0.59

Table 4.2. Mean ± standard deviation for all additional parameters investigated in a subset of age, gender, ethnicity, intraocular pressure, and central corneal thickness matched keratoconic patients ($n = 100$) and healthy participants ($n = 90$).

Parameters	Keratoconus	Healthy	P-value
Frequency of first harmonic (Hz)	375 ± 41.1	380 ± 46.0	0.50
Corneal deformation minimum for frequency over 150 Hz (μm)	−4.71 ± 9.05	−3.67 ± 4.35	0.31
Corneal deformation maximum for frequency over 150 Hz (μm)	15.9 ± 28.2	14.4 ± 28.1	0.70
'an' root mean square (mm)	0.15 ± 0.07	0.15 ± 0.04	0.29
Area under deformation amplitude curve (mm·ms)	13.9 ± 1.00	12.9 ± 1.41	< 0.01
Applanation length level (mm)	0.73 ± 0.07	0.78 ± 0.04	< 0.01
Deflection length level (mm)	6.31 ± 0.32	6.42 ± 0.32	0.02
Maximum corneal inward velocity (m s^{-1})	0.24 ± 0.22	0.21 ± 0.08	0.18
Maximum corneal deformation (mm)	0.88 ± 0.13	0.80 ± 0.09	< 0.01
Corneal energy dissipation (N m)	$(2.02 ± 0.65) \times 10^{-5}$	$(2.02 ± 0.65) \times 10^{-5}$	0.51
Transient stiffness coefficient (mN mm^{-1})	21.9 ± 2.55	22.9 ± 2.56	< 0.01
Energy absorbed area (mN·mm)	6.88 ± 1.32	6.37 ± 1.38	0.01
Ambrósio relational thickness to the horizontal profile (μm)	229 ± 124	473 ± 130	< 0.01
Stiffness parameter A1 (mmHg mm^{-1})	78.9 ± 19.4	106 ± 16.3	< 0.01
Corneal biomechanical index	0.88 ± 0.28	0.16 ± 0.28	< 0.01

Table 4.3. Area under the receiver operating analysis curve (AUC) ± standard deviation for the receiver operating characteristic analysis on the investigated algorithms that were significantly different between the subset of age, gender, ethnicity, intraocular pressure, and central corneal thickness matched keratoconic corneas ($n = 100$) and healthy corneas ($n = 90$).

Parameters	AUC
Corneal biomechanical index	0.93 ± 0.03
Ambrósio's relational thickness to the horizontal profile	0.91 ± 0.04
Applanation length level	0.75 ± 0.03
Maximum corneal deformation	0.69 ± 0.04
Area under deformation amplitude curve	0.63 ± 0.04
Energy absorbed area	0.61 ± 0.04
Transient stiffness coefficient	0.61 ± 0.04
Deflection length level	0.59 ± 0.05

P-values between receiver operating characteristic analysis curves

	AUDA	ALL	DLL	MCD	S_{TSC}	$A_{absorbed}$	ARTh
ALL	0.01	—					
DLL	0.58	0.02	—				
MCD	0.02	0.17	0.18	—			
S_{TSC}	0.50	< 0.01	0.84	0.04	—		
$A_{absorbed}$	0.67	< 0.01	0.75	0.13	0.96	—	
ARTh	< 0.01	0.03	< 0.01	0.02	< 0.01	< 0.01	—
CBI	< 0.01	< 0.01	< 0.01	< 0.01	< 0.01	< 0.01	0.31

Abbreviations: AUDA: area under the deformation amplitude curve, ALL: applanation length level, DLL: deflection length level, MCD: maximum corneal deformation, S_{TSC}: transient stiffness coefficient, $A_{absorbed}$: energy absorbed area, ARTh: Ambrósio's relational thickness to the horizontal profile, CBI: corneal biomechanical index.

specificity. The statistical differences between all the receiver operating characteristic analysis curves have been provided below.

4.4 Discussion

The standard parameters provided by the CST cannot be used in isolation to diagnose keratoconus. This limitation spurred researchers to derive additional parameters from the CST videos in the hope of improving the diagnostic value. However, these studies did not account for various factors such as age, intraocular pressure, and central corneal thickness—all of which are known to significantly influence corneal biomechanical properties. Nonetheless, most of these additional parameters were demonstrated to provide high levels of diagnostic ability, with some having an AUC of up to 0.98.

The current study compared fifteen novel parameters with each other in a single, large, cohort of keratoconic and healthy corneas carefully matched for age, gender, ethnicity, intraocular pressure, and central corneal thickness. Interestingly, a large number of the investigated parameters were still significantly different between healthy and keratoconic groups, even after controlling numerous factors. The statistical differences identified comparing healthy and keratoconus groups for the variables tested (except the 'energy absorbed area') may reflect previous *ex vivo* studies that have demonstrated a disparity in Young's modulus between keratoconic and healthy corneas. The increased extensibility of the cornea may be due to multiple factors, including the loss of stromal fibres, abnormal collagen fibre arrangement and orientation, a reduction in the number of crosslinks, and decreased stromal keratocyte density. The 'energy absorbed area' is a surrogate measurement for corneal viscoelasticity. This viscous behaviour primarily stems from the corneal ground substance—synthesised and regulated by corneal keratocytes that are abnormal in keratoconus. The defective keratocytes may alter ground substance composition and therefore cause an increased 'energy absorbed area'. The corneal biomechanical index was empirically derived through optimum parameter combination for the specific purpose of differentiating keratoconus from healthy corneas. Therefore, this parameter cannot be directly compared to a true physical parameter such as Young's modulus or hysteresis.

The corneal biomechanical index and Ambrósio's relational thickness to the horizontal profile had a reasonably high AUC (0.93 ± 0.03 and 0.91 ± 0.04, respectively) and either parameter can be used interchangeably as an adjunct to the diagnosis of keratoconus, as the AUCs did not differ significantly ($P = 0.31$) from each other. In comparison to previous studies, the AUCs for each parameter observed in this study vary widely. A possible explanation for the discrepancy is found in the definitions used to select the subset populations across studies. The current study specifically adjusted for multiple factors known to influence corneal biomechanics such as age, intraocular pressure, and central corneal thickness, in order to identify true differences between keratoconus and healthy corneas.

All of the parameters investigated focused on the anterior corneal curve, but it has been proposed that an increase in posterior corneal elevation may occur before any other signs of subclinical keratoconus. Li *et al* [19] recently proposed a new methodology for accurately deriving the posterior corneal location; however, further studies are required to determine if analysing the posterior corneal curve may provide greater diagnostic sensitivity.

Currently, tomographic evaluation of the cornea is the gold standard for diagnosing keratoconus. Further studies are needed to investigate the feasibility of combining tomographic evaluation of the cornea with biomechanical analysis to enable earlier detection of keratoconus.

References

[1] Tian L, Huang Y F, Wang L Q, Bai H, Wang Q, Jiang J J, Wu Y and Gao M 2014 Corneal biomechanical assessment using corneal visualization scheimpflug technology in keratoconic and normal eyes *J. Ophthalmol.* **2014** 147516
[2] Vellara H R, Ali N Q, Gokul A, Turuwhenua J, Patel D V and McGhee C N 2015 Quantitative analysis of corneal energy dissipation and corneal and orbital deformation in response to an air-pulse in healthy eyes *Invest. Ophthalmol. Vis. Sci.* **56** 6941–7

[3] Shetty R, Nuijts R M, Srivatsa P, Jayadev C, Pahuja N, Akkali M C and Sinha Roy A 2015 Understanding the correlation between tomographic and biomechanical severity of keratoconic corneas *Biomed. Res. Int.* **2015** 294197

[4] Koprowski R and Ambrosio R Jr 2015 Quantitative assessment of corneal vibrations during intraocular pressure measurement with the air-puff method in patients with keratoconus *Comput. Biol. Med.* **66** 170–8

[5] Wang L K, Tian L and Zheng Y P 2016 Determining *in vivo* elasticity and viscosity with dynamic Scheimpflug imaging analysis in keratoconic and healthy eyes *J. Biophotonics* **9** 454–63

[6] Steinberg J, Katz T, Lucke K, Frings A, Druchkiv V and Linke S J 2015 Screening for keratoconus with new dynamic biomechanical *in vivo* Scheimpflug analyses *Cornea* **34** 1404–12

[7] Vinciguerra R, Ambrosio R Jr, Elsheikh A, Roberts C J, Lopes B, Morenghi E, Azzolini C and Vinciguerra P 2016 Detection of keratoconus with a new biomechanical index *J. Refract. Surg.* **32** 803–10

[8] Roberts C J, Mahmoud A M, Bons J P, Hossain A, Elsheikh A, Vinciguerra R, Vinciguerra P and Ambrosio R Jr 2017 Introduction of two novel stiffness parameters and interpretation of air puff-induced biomechanical deformation parameters with a dynamic Scheimpflug analyzer *J. Refract. Surg.* **33** 266–73

[9] Rabinowitz Y S 1998 Keratoconus *Surv. Ophthalmol.* **42** 297–319

[10] Zadnik K, Barr J T, Edrington T B, Everett D F, Jameson M, McMahon T T, Shin J A, Sterling J L, Wagner H and Gordon M O 1998 Baseline findings in the Collaborative Longitudinal Evaluation of Keratoconus (CLEK) study *Invest. Ophthalmol. Vis. Sci.* **39** 2537–46

[11] Koprowski R, Kasprzak H and Wróbel Z 2014 New automatic method for analysis and correction of image data from the corvis tonometer *Comput. Methods Biomech. Biomed. Eng.: Imaging Vis.* **5** 27–35

[12] Andreassen T T, Simonsen A H and Oxlund H 1980 Biomechanical properties of keratoconus and normal corneas *Exp. Eye Res.* **31** 435–41

[13] Nash I S, Greene P R and Foster C S 1982 Comparison of mechanical properties of keratoconus and normal corneas *Exp. Eye Res.* **35** 413–24

[14] Vellara H R and Patel D V 2015 Biomechanical properties of the keratoconic cornea: a review *Clin. Exp. Optom.* **98** 31–8

[15] Sherwin T and Brookes N H 2004 Morphological changes in keratoconus: pathology or pathogenesis *Clin. Exp. Ophthalmol.* **32** 211–7

[16] Ali N Q, Patel D V and McGhee C N 2014 Biomechanical responses of healthy and keratoconic corneas measured using a noncontact scheimpflug-based tonometer *Invest. Ophthalmol. Vis. Sci.* **55** 3651–9

[17] Swets J A 1988 Measuring the accuracy of diagnostic systems *Science* **240** 1285–93

[18] Robin X, Turck N, Hainard A, Tiberti N, Lisacek F, Sanchez J C and Muller M 2011 pROC: an open-source package for R and S+ to analyze and compare ROC curves *BMC Bioinform.* **12** 77

[19] Li T, Tian L, Wang L, Hon Y, Lam A K, Huang Y, Wang Y and Zheng Y 2015 Correction on the distortion of Scheimpflug imaging for dynamic central corneal thickness *J. Biomed. Opt.* **20** 56006

IOP Publishing

Air-Puff Tonometers
Challenges and insights
Robert Koprowski

Chapter 5

Intraocular pressure and three-dimensional corneal biomechanics

Pablo Pérez-Merino, Nandor Bekesi and Antonio Fernández-López

Glaucoma is the main cause of irreversible blindness worldwide and its prevalence is increasing exponentially with age, so nearly 80 million people in the world will be diagnosed with glaucoma by 2020 [1]. It is a complex set of age-related multifactorial conditions characterized by a progressive disease of the optic nerve. The underlying mechanisms are by no means fully understood, but several glaucoma theories rely on the fluid flow-related phenomena, in which impairment in the aqueous humor flow is often associated with elevated intraocular pressure (IOP).

IOP is a function of the rate at which the aqueous humor enters the eye (inflow) and the rate at which it leaves the eye (outflow). When inflow equals outflow, a steady state exists and the IOP remains fairly constant. However, under non-physiological conditions the dynamic equilibrium and the delicate pressure-regulating mechanisms fail, leading to an unidentified outflow resistance [2–4].

Although high IOP is certainly not the only factor contributing to optic nerve damage and visual field loss, an increase in IOP is observed in most glaucoma patients and is the only risk factor that can be clinically modified [5]. However, the problem is that an IOP measurement is just a one-time recording with low sensitivity (less than 30%) [6] and its accuracy is influenced by the *dynamic behavior of the intraocular pressure* and multiple *corneal factors* (e.g. patients with thicker corneas show a higher value of IOP than actually exists) [7–9]. Besides, the corneal structure in keratoconus [10] or post-LASIK [11] corneas affects typical intraocular pressure measurements since these cases showed marked irregularities in both corneal surfaces and progressive corneal thinning (in keratoconus), and reduction of thickness and curvature (in post-LASIK corneas), and therefore have different material properties.

doi:10.1088/2053-2563/aafee5ch5

5.1 Behind the need for change in glaucoma diagnosis: IOP and corneal biomechanics

A consensus statement from the World Glaucoma Association about screening is that '*a screening test should be safe, easy to administer and interpret, portable, quick, acceptable to the people who are tested, able to obtain results in the majority of tested individuals and sufficiently valid to distinguish between those who do and those who do not have glaucoma.* [12]'

Glaucoma screening during routine examination has been traditionally attempted by: IOP measurement, optic nerve check, irido-corneal angle quantification and visual field-testing. However, IOP has not been proven to be a reliable tool for glaucoma diagnosis since there is no cut-off value of IOP that provides the necessary balance between sensitivity and specificity. IOP measurements showed a sensitivity of only 22.1%. Computerized visual field screening showed higher sensitivity (58.1%), but it is far from ideal since nearly half of the patients with glaucoma were not positively identified and the disease progressed irreversibly affecting different areas of the visual field. Unluckily, the automated imaging techniques did not achieve the Prevent Blindness America criteria for minimum performance of a screening test ($\geqslant$ 95%, specificity; $\geqslant$ 85%, sensitivity) [6, 12, 13].

Tonometers measure IOP by determining the resistance of the cornea to an applied load, and in a simplistic description, most tonometry devices consider the cornea to be a homogeneous structure with an average corneal thickness in IOP readings. However, the cornea has particular *shape patterns* (e.g. astigmatism and irregularities in both corneal surfaces) [14], and it is a complex structure with marked *anisotropy* (e.g. the properties are not directionally uniform) [15] and *heterogeneity* (e.g. the characteristics are determined by the interaction of the collagen fibril structure with the extracellular matrix and other components) [16]. Thus, it manifests different biomechanical properties when tension is applied in different directions and follows the typical features of soft tissues: it is a viscoelastic material with non-linear elastic behavior and has a complex structure via a different stress–strain relationship that depends on the loading rates, velocity and time [17–19].

Corneal biomechanics is strongly intertwined with the concept of (i) *elasticity*, that represents the ratio between the applied force and the magnitude of corneal deformation, and (ii) *viscoelasticity*, which describes the time-dependent nature of the cornea; therefore, the response will be a function of the velocity and time of the applied force, and will be characterized by the corneal deformation and recovery to the initial shape.

Therefore, adding *patient-specific geometry* and *corneal biomechanical properties in three dimensions* to IOP readings (i) will refine IOP measurements, (ii) will increase the sensitivity and specificity in screening, and (iii) will improve the diagnosis and treatment of the disease.

5.2 Patient-specific geometry: three-dimensional corneal shape

The characteristic shape pattern of corneal surfaces is commonly assessed in the clinic by means of corneal topography. To date, there is a huge variety of commercial systems to measure corneal topography that can be classified based on the imaging principle: specular reflection, scattered light, Scheimpflug imaging and optical coherence tomography (OCT) [20]. Scheimpflug and OCT are the only imaging techniques that generate true elevation points with micron-resolution of both corneal surfaces, anterior and posterior [21–24]. Figure 5.1 shows an illustration of three-dimensional corneal representation from OCT images.

Assuming that the corneal surface is given by $z = f(x, y)$ in a Cartesian system with first and second derivatives continuous at any point, there are three ways of representing corneal topography:

- By *surface elevation* $f(x, y)$ with respect to a reference surface (e.g. sphere). A typical reference sphere is one with the minimum standard deviation with respect to the corneal surface and with the same optical axis. The best-fit sphere to evaluate the topography of the cornea is calculated using a least-squares method.
- By *local slopes* with respect to the reference sphere, since at any point on the surface the slope is a function of the direction.
- By *local curvature*, for a given point there is a maximum value in a certain direction and a minimum value in the perpendicular direction.

The corneal surface data is commonly expressed in Euclidean coordinates and it is fitted by standard functions: sphere, ellipsoid, conicoid, biconocoid or Zernike polynomial expansions (note that these are fits to surface elevations, not corneal wave aberrations). The corneal thickness is calculated by direct subtraction of the posterior surface from the anterior surface in the measured area.

Corneal topography and thickness can be measured easily with commercial systems, but strategies to decode corneal biomechanics are one of the major challenges.

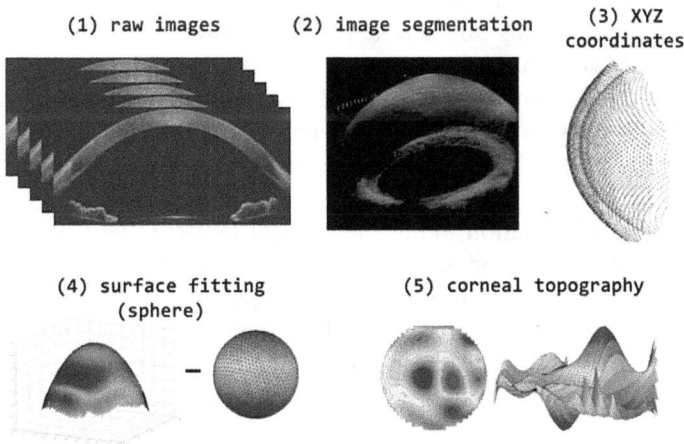

Figure 5.1. Illustration of the OCT segmentation process and calculation of the topographic map from direct subtraction of the elevation data minus the best-fitted sphere.

5.3 Corneal biomechanics: traditional *ex vivo* mechanical testing

Histologically, the cornea is an inhomogeneous cellular and fibrillar structure composed of five layers: epithelium, Bowman's layer, stroma, Descemet's membrane and endothelium. The cornea mainly contains water (78%), regularly arranged collagen fibrils, proteoglycans and keratocytes [14, 25].

The bulk of the cornea is formed by the *stroma* (approximately 90% of corneal thickness), which in the human adult is approximately 450–550 µm thick centrally, 550–750 µm in the periphery and is composed predominantly of flattened and stacked collagenous lamellae (200–250 layers). Stromal collagen assembles to form long fibrils that in the human cornea show a uniform diameter of approximately 31–34 nm, and are separated by 20–50 nm. Collagen fibers are approximately 1–2 µm thick, 10–200 µm wide and are generally organized into independent bundles, lamellae, which are embedded in an extracellular matrix. The microstructure of the stroma is highly heterogeneous, since depending on the specific region the collagen fibers are more closely packed, less hydrated or show stronger junctions. This particular arrangement maintains an ordered transparent structure while also enhancing mechanical strength [26–30].

In *ex vivo* human corneas, x-ray scattering and scanning electron microscopy measurements reveal that collagen fibers have a preferred orientation, distribution and organization in the stromal structure, with lamellae oriented in the inferior–superior and nasal–temporal directions around the center of the cornea in the central-posterior stroma, radial preferred orientation within 2 mm of the limbus and a disorganized orientation structure in the anterior part of the stroma, with the presence of a higher interweaving and branching of collagen fibers in the anterior cornea compared to the posterior (showing × 4 times higher branching density for the region below the Bowman's layer). These characteristics show the cornea to be an anisotropic, non-linear and inhomogeneous material with a complex behavior, therefore exhibiting different mechanical properties [15, 17, 31–35].

Most of the traditional studies on corneal biomechanics have been conducted on *ex vivo* corneas employing uniaxial tension measurements or pressure inflation testing combined with image registration. Conventional laboratory experiments have been performed through strip uniaxial extensiometry, where the corneas were isolated from the enucleated eye and different corneal strips of around 1–2 mm width were clamped to the extensiometer. However, the main limitation of the extensiometry technique is the disruption of the fibril distribution, which clearly modifies the stress distribution along the corneal tissue. Other methods proposed corneo-scleral shell inflation on an artificial pressure chamber or whole ocular globe inflation to mimic the natural intraocular environment. These inflation techniques aim to control the intraocular pressure and overcome the limitations of extensiometry, but post-mortem time and experimental conditions (humidity, temperature and storage solution) influence the hydration state and the organized microstructure of the cornea.

In order to estimate the biomechanical parameters, these methods require controlled mechanical forces to be applied to the cornea through contact: typically a mechanical force along one axis in extensiometry [19, 36] or a radial mechanical force by increasing and decreasing the intraocular pressure in the inflation techniques [37, 38]. The displacements and strains of the cornea are expected to be zero as no stress is applied. With increasing loads the strain is measured by corneal elongation in extensiometry or by measuring different landmarks of the corneal surface in the inflation techniques (e.g. apex displacement or changes in the radius of curvature), extracting different material properties:

- the *ocular rigidity coefficient*, by analyzing the slope of the pressure–volume curve;
- *Young's modulus (elastic) and hysteresis (viscoelastic)*, by evaluating the different mechanical responses of the stress–strain curve. The elastic modulus defines the proportionality between the magnitude of corneal deformation and the stress; whereas the viscoelastic components describe the time-dependent relationship between corneal deformation and stress.

However, the major downside of these measurement techniques (axial extensiometry or inflation testing) is that they are invasive in nature, the *in vivo* conditions cannot be fully reproduced and, as a result, the corneal elasticity modulus reported in the literature showed wide variation (up to two orders of magnitude: ranging from 0.1 to 57 MPa [39]).

5.4 Corneal biomechanics and IOP: linking non-invasive imaging and simulations

Different strategies focused on new paradigms for accurate analysis of experimental inflation testing; however, the practical use of corneal biomechanical models have been limited by the reliability of the input data, the reproducibility of the measurement conditions and the validation of the methodology. Therefore, the biomechanical models will largely benefit from accurate and non-invasive *in vivo* measurements.

Some existing commercial methods to measure the corneal biomechanical properties and IOP *in vivo* are based on a corneal air puff coupled with a sensor: (1) photodetector (ORA, by Reichert Ophthalmic Instruments Inc, Depew, NY) or (2) Scheimpflug imaging (Corvis ST, by Oculus Optikgeräte GmbH, Wetzlar, Germany). ORA measures IOP and corneal mechanical properties (hysteresis and rigidity) using a dynamic bidirectional applanation process, in which changes in reflectivity are captured as a rapid pulse of air deforms the cornea. Corvis ST provides two-dimensional cross-section images of the deforming cornea during applanation and measures two positional parameters: apex displacement as a function of time (temporal corneal deformation) and the cross-section of the deformed shape of the cornea at maximum concavity (spatial corneal deformation).

The ORA derives values of the inward and outward pressure obtained during dynamic applanation, which as a result of the viscous damping produces a delayed response. This difference is estimated as corneal hysteresis. Another biomechanical parameter related to the elastic properties is also described as the corneal resistance factor. Apart from corneal biomechanical parameters, the system was also presented with the capability of measuring IOP readings less dependent on corneal thickness than the applanation tonometers. However, the signal provided by the ORA is based on changes in light intensity captured by a photodetector (not based on corneal imaging) and does not provide a direct measurement of corneal deformation upon applanation. Due to the indirect nature of the measurements, different studies in the literature questioned the sensitivity of the technique to register changes of the parameters associated with corneal hysteresis in normal and keratoconic corneas [40–42].

The Corvis provides corneal imaging in a cross-section during the applanation process (acquiring 140 horizontal cross-sectional images during the 30 ms deformation event; the speed property of the high-speed camera is about 4330 fps), so the system records different dynamic phases of the cornea in two dimensions in real time: (i) deformation amplitude, (ii) applanation length and time, (iii) corneal velocity and (iv) the recovery position. In addition, the Corvis also measures the corneal thickness and IOP [43, 44].

The common exported Corvis parameters from the deformation data are: (i) peak-to-peak distance (lateral distance between two convex peaks at the maximal deformation), (ii) temporal symmetry factor (symmetry of the corneal apex *vs.* time), and (iii) corneal deformation as a function of time: elastic air-pulse response and viscoelastic recovery. The viscoelastic component of the cornea was almost neglected with the impact of a very fast air pulse, resulting in a quasi-elastic response during the stress. Corneal hysteresis is only observed during the recovery phase [45, 46].

The spatial and temporal deformation Corvis profiles could be analyzed by using computational fluid dynamics (CFD) simulation and inverse optimization methods to retrieve the biomechanical parameters that better match the experimental conditions [45, 47]. Besides, some intrinsic parameters and assumptions as input data need to be considered:

- the *corneal geometry* and *corneal thickness;*
- the *intraocular pressure* and *boundary conditions;*
- the *microstructural composition of the corneo-scleral mesh* (e.g. distribution of collagen fibers);
- the intrinsic characteristics of the *air pulse.*

Some simple elastic mechanical models with real data were reported in earlier publications [48]; although they showed valuable information, critical assumptions limited the applicability of the results: (i) linear elasticity is only present within a very small strain range, (ii) the constituent materials of the cornea (collagen fibers and extracellular matrix) showed a nearly incompressible, anisotropic and viscoelastic behavior, in which the extracellular matrix and collagen fibers are generally uncoupled and their alteration could be related to a viscoelastic failure.

To overcome these limitations and describe the material properties at higher stress ranges, *non-linear elastic, hyperelastic* and *viscoelastic models* have been incorporated into computer models [18, 49–52]. Different commercial software

(e.g. ANSYS, COMSOL, ABAQUS) is commonly used to propose theoretical optimization methods of corneal deformation as close to reality as possible. But some parameters are needed to replicate the experiments (figure 5.2):
  - a detailed geometrical model and a suitable reference;
  - the description of the material constituents of the modeled structure;
  - the specifications of the boundary conditions;
  - the operative forces.

Likewise, the validation and verification of the numerical experiments should include the accuracy of the method to fit the experimental results, adopted assumptions and possible source of errors, order of complexity of the models and the uncertainties of the input data.

The *corneal geometry* and *corneal thickness* can be considered in two dimensions or in three dimensions:
  - in *two dimensions*, by assuming the cornea as a part of a spherical shell with axial symmetry and thickness increasing linearly along the meridional direction;
  - in *three dimensions*, by incorporating in a quadrilateral mesh of a node–element structure the patient-specific irregularities of the topographic pattern and (for example the XYZ coordinates or the Zernike coefficients), the

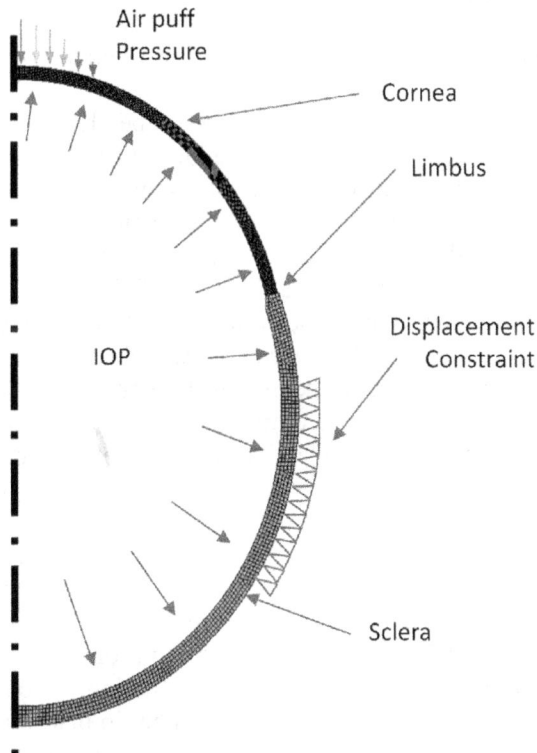

Figure 5.2. Finite element mesh of the eye with the applied loads and boundary conditions. Reproduced from [53], CC BY-SA 4.0.

transitional zone between the cornea and the sclera (coordinates of the limbus junction and node description) and the geometrical data of the sclera (the sclera plays a secondary role since the cornea receives the impact of the air pulse). The nodes and elements represent the connecting points and the small units of the corneo-scleral mesh, respectively.

The *intraocular pressure* could be defined as a pressure load assumed to be equally distributed in the posterior corneal area. Besides, more realistic models incorporate the fluid–structure interaction state in IOP modeling, since a uniform pressure distribution for IOP definition is not valid when a compression load impacts the cornea (e.g. the air pulse).

The *microstructural composition of the corneo-scleral mesh* is usually described with specific material properties in order to compute the simulations under defined boundary conditions as forces or displacements. The stroma is highly heterogeneous in the central to peripheral and anterior to posterior dimensions, so the form of the strain of the cornea could be modeled with:

- different material behaviors (e.g. linear viscoelastic material or hyperelastic Mooney–Rivlin material);
- formulation of stress/strain/deformation tensors;
- preferred material directions;
- assumptions in the elastic modulus, Poisson's ratio, shear modulus or other material parameters included in the constitutive equation.

Prony series and relaxation time constants could be also applied to describe the viscoelastic behavior of the system. Further refinement of the model is possible by adding the angle and distribution of the fibers in the material configuration: nasal–temporal and superior–inferior orientation in the cornea and single circumferential orientation in the limbus. Limbus material parameters are commonly assumed to be identical as for the cornea, but with different preferential directions. The sclera could be assumed to be an isotropic hyperelastic material with symmetry boundary conditions. The scleral material is modeled with higher stiffness, in comparison with the cornea. Due to the high water content, the corneo-scleral shell is usually considered to be a nearly incompressible material. Finally, the number of nodes and elements of the corneo-scleral mesh will define the complexity of the system and the computational time.

The *Corvis air-pulse parameters are*: pressure (peak pressure of 15 kPa, approximately 115 mmHg—×8 the normal value of intraocular pressure), spatial (an air speed of 115 m s^{-1} impacts on the corneal surface and produces a circular deformation on the central area of 3 mm; central diameter is around 2.4 mm) and temporal distribution (27.5 ms duration; although only 20.63 ms contributed effectively to the corneal deformation). In the simulation, a collimated pressure is applied on the anterior corneal surface (with the tube tip placed at 11 mm from the eye), and the impact varied as a function of time, distance and location to the apex and corneal shape. Figure 5.3 illustrates the temporal air-pulse profile of the Corvis system.

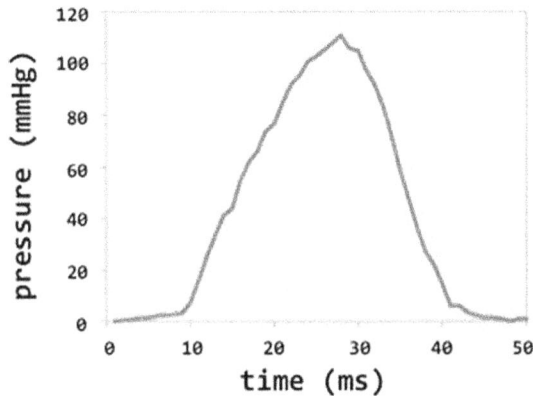

Figure 5.3. Air-pulse characterization: experimentally measured temporal air-pulse profile. Reproduced from [45], CC BY-SA 4.0. © 2014 Kling *et al.*

The pressure distribution of the air pulse was characterized by measuring the pressure change over time at the center of the air exit and by modeling the setup (air pulse + corneal impact) in CFD simulations. The deformed shape of the cornea can significantly alter the pressure distribution at the corneal surface, so the deformation of the cornea should be considered in determining the air-pulse characteristics (figure 5.4) [45].

When the cornea receives the impact of the air pulse, it experiences a bending state of stress: with changes in the stress state between the anterior and posterior stroma (the anterior stroma works in compression while the posterior stroma works in tension). With these premises, different studies have demonstrated a strong correlation between corneal rigidity and corneal deformation, showing the following parameters high contribution to resistance: corneal thickness, intraocular pressure, material parameters and corneal geometry (from higher to lower contribution) [47, 54–57]. Roberts (2014) [58] also showed that IOP is the strongest biomarker of corneal deformation amplitude followed by corneal stiffness. Kling and Marcos (2013) [59] and Ariza-Gracia *et al* (2015) [47] found that the maximum displacement of the corneal apex due to the air-pulse impact varies linearly with IOP for different corneal stiffness, and showed that different combinations of corneal stiffness and IOP could result in the same corneal deformation, describing overlapping between them. So it could not be possible to differentiate between IOP and the corneal material properties without knowing their individual parameters.

Therefore, for an accurate IOP reading, reconstruction of the individual structural properties of the cornea should be considered. Figure 5.5 illustrates the analysis of the IOP and the mechanical properties of the cornea using a multivariable optimization process. Besides, the necessity of fluid–structure interaction (using the ocular humors as a pressurized fluid with mass) in the simulations was also pointed out to accurately reproduce the effect of the air pulse in the reconstruction of the intraocular pressure [60], since considering the aqueous/vitreous humors as a fluid could modify the final value of the IOP.

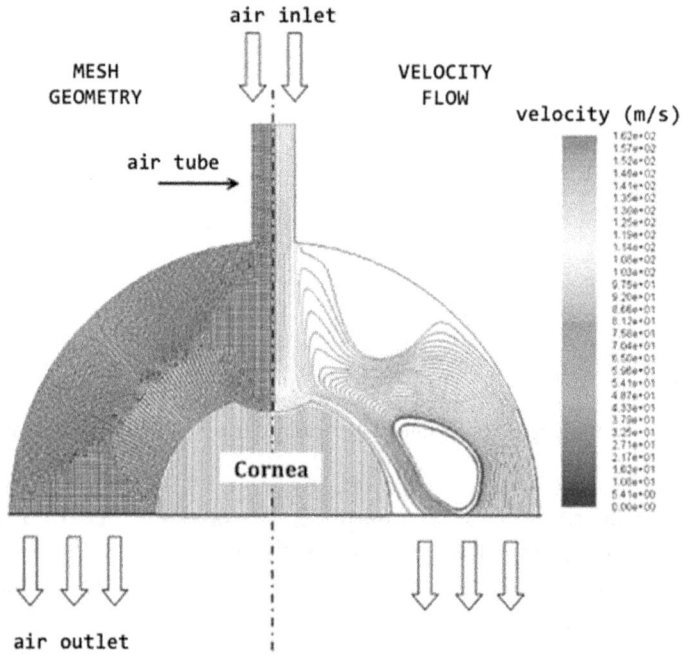

Figure 5.4. Geometry model of the cornea at maximal deformation: mesh of cells of the modeled air volume (left) and streamlines (right) colored by the flow velocity distribution. Reproduced from [45], CC BY-SA 4.0. © 2014 Kling *et al.*

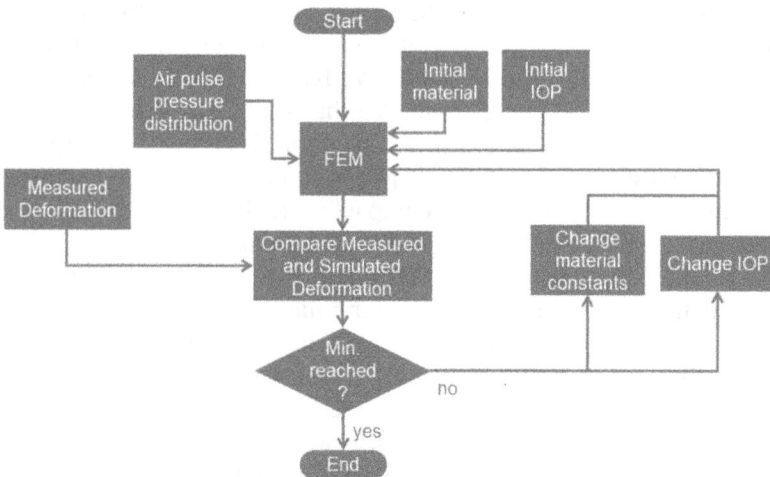

Figure 5.5. Block diagram of the iteration process to determine the corneal material properties and the IOP by finite element simulations.

5.5 Novel concepts to decipher corneal biomechanics in 3D and IOP

The results of numerical methods have explained the relationship between corneal biomechanics and deformation, but the reliability of theoretical studies is strongly dependent on the input data needed to build the routines for simulations: *ex vivo* results from extensiometry or inflation testing, two-dimensional measurements based on air-pulse measurements (Corvis) or material properties of the collagen fibers analyzed from x-ray scattering or scanning electron microscopy. Furthermore, assumptions on the boundary conditions and material constitutive equations could also affect the final accuracy and validity of the simulations.

A recent experimental study reported that IOP readings were very different between ORA, Corvis ST and tonometry [61]. ORA and Corvis ST show one-dimensional (temporal) and two-dimensional (spatio-temporal) deformation of the cornea, thus, with no doubt, a final adjustment for *three-dimensional corneal mechanical properties* is of crucial importance for real and true IOP measurements.

Different experimental imaging-based systems are currently under investigation to measure corneal biomechanics in three dimensions and overcome the lack of more direct measurements of corneal deformation *in vivo*: optical coherence tomography (OCT), ultrasound, interferometric and Brillouin scattering techniques.

The three-dimensional scanning architecture, high resolution and high speed of the OCT presents several advantages over Scheimpflug imaging, being the swept source configuration of the latest milestone. OCT has the following important advantages to detect the dynamic phenomena with micron-resolution: (i) the laser source is usually infrared, so, with controlled power, it is not harmful to human tissue and is comfortable for the patient; (ii) the system is based on a low-coherence interferometer and the resolution is limited by the coherence length of the laser, thus high resolution (1–10 µm) can be achieved; (iii) the system can be fiber based, therefore it could be easily made compact and low cost; (iv) real-time imaging can be achieved; and (v) higher speed over other imaging technologies. Different modalities of OCT synchronized with mechanical corneal excitation systems have been used to register the dynamics of the cornea, a corneal air puff being the most common mechanical excitation system [62, 63].

The addition of elastographic contrast to OCT in recent studies has improved the ability to differentiate the composition and structure of the cornea (up to the subpixel resolution), providing therefore higher strain measurement accuracy of the cornea [64, 65]. There are two main categories of optical coherence elastography: phase-sensitive and speckle tracking, both techniques analyze the displacement amplitude attenuation, elastic wave speed, dispersion of the elastic waves and natural frequency of the vibrations of the focused air pulse in the corneal surface.

The assessment of corneal biomechanical properties has been also proposed through other non-invasive methods. *Ultrasound elastography* measures corneal elasticity by analyzing the shear wave (as shear waves can induce mechanical oscillations in a limited local region), though the corneal strain and the hydration of the cornea typically affect the wave speed. In addition, the technique is limited by low resolution and the correlation between the corneal acoustic and the elastic properties has been only demonstrated for low strain loads [66–68].

High-speed image-based strain measurements could directly track corneal surface markers in three dimensions during deformation to inversely calculate displacements and strain in a non-invasive approach. *High rate interferometric imaging* has been recently proposed to characterize the three-dimensional corneal biomechanical properties; but it shows limited deformation range and it is too sensitive to small displacements [69]. *Digital image correlation (DIC)* is widely accepted for mechanical testing and includes the tracking of the same points or pixels of images before and after deformation; when DIC is applied to continuum mechanics, strain can be simultaneously obtained with displacements by using speckle tracking and cross-correlation algorithms [70]. However, a non-invasive speckle pattern is required for an *in vivo* application. High rate interferometric and DIC could be suitable techniques for testing in three dimensions with a large field of view and micro-displacement accuracy: softening, creep, relaxation, rigidity and hysteresis.

Brillouin light scattering arises from the interaction of light and sound waves that are inherently present in the tissues and allows non-invasive measurements of the biomechanical properties. The last configuration of the technique combines a confocal microscope with a spectrometer. However, although Brillouin is the only methodology that has demonstrated a full three-dimensional spatial mapping of the corneal biomechanical properties *in vivo* [71], its potential application in ophthalmology had not been fully explored because to date it requires longer acquisition times and does not provide absolute inherent mechanical parameters (Brillouin modulus is still under investigation).

Finally, air-pulse devices have become a reference to study the effect of corneal deformation *in vivo*. However, an air pulse is unpleasant for the patient and the large corneal deformation response (by more than 1 mm, being quite large as the corneal thickness is under 500 µm) of the commercial devices produces large displacement of intraocular structures (aqueous humor, iris and lens), adjacent structures (sclera) and ocular motion into the orbit, resulting in large global deformation of the eye (figure 5.6). Other excitation systems involve acoustic [72], mechanical [73] or photothermal [74] stimulation, but some limitations are still associated with these such as: the influence from surrounding tissues (sclera or aqueous) in acoustic, the tissue compression in mechanical, and the light absorption or the safe level of heat in photothermal. To meet these limitations a corneo-scleral focused micro air-pulse system capable of inducing a surface wave in the cornea has been demonstrated in OCT imaging [75].

5.6 Looking for new alternatives in IOP measurements

The current commercial technologies including clinical diagnosis may be sufficiently sensitive to manage glaucoma. However, to date, clinical grading is not without problems, since it is prone to false positives and false negatives due to the lack of a cut-off value, it misses most cases of subclinical stages and there is a real lack of critical characteristics for an evidence-based medicine: objective, quantitative, non-

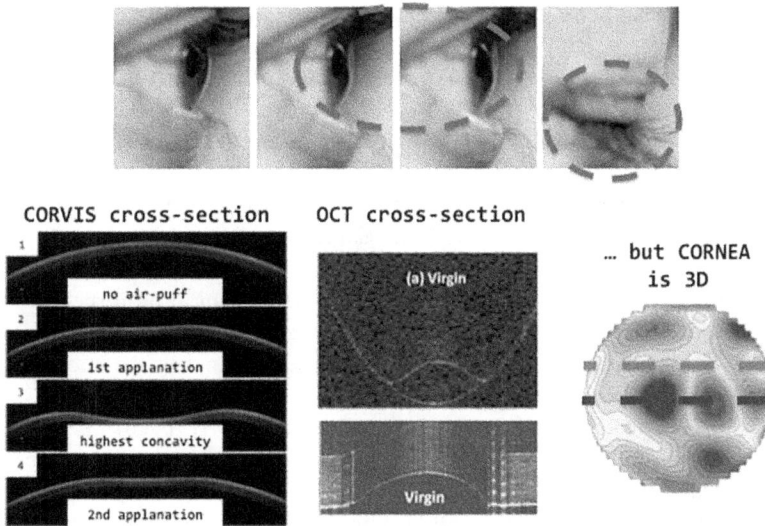

Figure 5.6. (Top) Illustration of the large corneal deformation response, movement of the eye into the orbit and protective blink during the applanation air-pulse process (images adapted from https://research.qut.edu.au/clvol/research/imaging-of-the-eye/). (Bottom) Corneal cross-section images (Corvis and OCT) during the applanation process and representation of a corneal topography to illustrate that the response and the biomechanical parameters would be affected by regional differences of the measured meridian. Reproduced from [63] © 2012 Optical Society of America.

invasive, three-dimensional and dynamic. Despite advanced experimental methods and simulations, the *in vivo* mechanical response of the cornea, and therefore the real IOP value, still remains unknown.

The medical community is in the process of wide-ranging changes at the moment due to the large ecosystem in imaging technologies. High-speed sensors and three-dimensional (3D) spatial motion tracking will allow us to work on revealing new features in biomedicine, biomechanics and microfluidics, offering both remarkable opportunities and challenges in eye care. In particular, tracking the fluid flow-related phenomena and the material corneo-sclera shell deformation in micro scale is fundamental for understanding the underlying mechanisms in the worldwide leading causes of visual impairment. Corneal biomechanics, such as elasticity and hysteresis, might be early diagnostic biomarkers in keratoconus, but also have substantial and wide influence on IOP measurement, of crucial importance for the screening and management of glaucoma.

References

[1] Tham Y C, Li X, Wong T Y, Quigley H A, Aung T and Cheng C Y 2014 Global prevalence of glaucoma and projections of glaucoma burden through 2040: a systematic review and meta-analysis *Ophthalmology* **121** 2081–90

[2] Fitt A D and Gonzalez G 2006 Fluid mechanics of the human eye: aqueous humour flow in the anterior chamber *Bull. Math. Biol.* **68** 53–71

[3] Johnson M 2006 What controls aqueous humour outflow resistance? *Exp. Eye Res.* **82** 545–57

[4] Tamm E R 2009 The trabecular meshwork outflow pathways: structural and functional aspects *Exp. Eye Res.* **88** 648–55

[5] Marquis R E and Whitson J T 2005 Management of glaucoma: focus on pharmacological therapy *Drugs Aging* **22** 1–21

[6] Salim S, Netland P A, Fung K H, Smith M E and Aldridge A 2009 Assessment of the Student Sight Savers Program methods for glaucoma screening *Ophthalmic Epidemiol.* **16** 238–42

[7] Liu J and Roberts C J 2005 Influence of corneal biomechanical properties on intraocular pressure measurement: quantitative analysis *J. Cataract Refract. Surg.* **31** 146–55

[8] Sultan M B, Mansberger S L and Lee P P 2009 Understanding the importance of IOP variables in glaucoma: a systematic review *Surv. Ophthalmol.* **54** 643–62

[9] Broman A T, Congdon N G, Bandeen-Roche K and Quigley H A 2007 Influence of corneal structure, corneal responsiveness, and other ocular parameters on tonometric measurement of intraocular pressure *J. Glaucoma* **16** 581–8

[10] Firat P G, Orman G, Doganay S and Demirel S 2013 Influence of corneal parameters in keratoconus on IOP readings obtained with different tonometers *Clin. Exp. Optom.* **96** 233–7

[11] Chang D H and Stulting R D 2005 Change in intraocular pressure measurements after LASIK the effect of the refractive correction and the lamellar flap *Ophthalmology* **112** 1009–16

[12] Taylor H 2011 Glaucoma screening in the real world *Ophthalmology* **118** 1008 author reply 1008–9, 1009–10

[13] Maul E A and Jampel H D 2010 Glaucoma screening in the real world *Ophthalmology* **117** 1665–6

[14] Atchison D A and Smith G 2000 *Optics of the Human Eye* (Oxford: Butterworth-Heinemann)

[15] Meek K M and Boote C 2009 The use of X-ray scattering techniques to quantify the orientation and distribution of collagen in the corneal stroma *Prog. Retin. Eye Res.* **28** 369–92

[16] Winkler M, Chai D, Kriling S, Nien C J, Brown D J, Jester B, Juhasz T and Jester J V 2011 Nonlinear optical macroscopic assessment of 3-D corneal collagen organization and axial biomechanics *Invest. Ophthalmol. Vis. Sci.* **52** 8818–27

[17] Boote C, Dennis S, Huang Y, Quantock A J and Meek K M 2005 Lamellar orientation in human cornea in relation to mechanical properties *J. Struct. Biol.* **149** 1–6

[18] Pandolfi A and Manganiello F 2006 A model for the human cornea: constitutive formulation and numerical analysis *Biomech. Model. Mechanobiol.* **5** 237–46

[19] Elsheikh A and Anderson K 2005 Comparative study of corneal strip extensometry and inflation tests *J. R. Soc. Interface* **2** 177–85

[20] Mejia-Barbosa Y and Malacara-Hernandez D 2001 A review of methods for measuring corneal topography *Optom. Vis. Sci.* **78** 240–53

[21] Ortiz S, Siedlecki D, Perez-Merino P, Chia N, de Castro A, Szkulmowski M, Wojtkowski M and Marcos S 2011 Corneal topography from spectral optical coherence tomography (sOCT) *Biomed. Opt. Express* **2** 3232–47

[22] Perez-Merino P, Ortiz S, Alejandre N, de Castro A, Jimenez-Alfaro I and Marcos S 2014 Ocular and optical coherence tomography-based corneal aberrometry in keratoconic eyes treated by intracorneal ring segments *Am. J. Ophthalmol.* **157** 116–27.e1

[23] Perez-Merino P, Ortiz S, Alejandre N, Jimenez-Alfaro I and Marcos S 2013 Quantitative OCT-based longitudinal evaluation of intracorneal ring segment implantation in keratoconus *Invest. Ophthalmol. Vis. Sci.* **54** 6040–51

[24] Ortiz S, Siedlecki D, Grulkowski I, Remon L, Pascual D, Wojtkowski M and Marcos S 2010 Optical distortion correction in optical coherence tomography for quantitative ocular anterior segment by three-dimensional imaging *Opt. Express* **18** 2782–96

[25] Knupp C, Pinali C, Lewis P N, Parfitt G J, Young R D, Meek K M and Quantock A J 2009 The architecture of the cornea and structural basis of its transparency *Adv. Protein Chem. Struct. Biol.* **78** 25–49

[26] Morishige N, Takagi Y, Chikama T, Takahara A and Nishida T 2011 Three-dimensional analysis of collagen lamellae in the anterior stroma of the human cornea visualized by second harmonic generation imaging microscopy *Invest. Ophthalmol. Vis. Sci.* **52** 911–5

[27] Polack F M 1961 Morphology of the cornea. I. study with silver stains *Am. J. Ophthalmol.* **51** 1051–6

[28] Jester J V 2008 Corneal crystallins and the development of cellular transparency *Semin. Cell Dev. Biol.* **19** 82–93

[29] Meek K M and Knupp C 2015 Corneal structure and transparency *Prog. Retin. Eye Res.* **49** 1–16

[30] Hjortdal J O 1996 Regional elastic performance of the human cornea *J. Biomech.* **29** 931–42

[31] Meek K M and Newton R H 1999 Organization of collagen fibrils in the corneal stroma in relation to mechanical properties and surgical practice *Refract. Surg.* **15** 695–9

[32] Meek K M, Tuft S J, Huang Y, Gill P S, Hayes S, Newton R H and Bron A J 2005 Changes in collagen orientation and distribution in keratoconus corneas *Invest. Ophthalmol. Vis. Sci.* **46** 1948–56

[33] Hayes S, Boote C, Lewis J, Sheppard J, Abahussin M, Quantock A J, Purslow C, Votruba M and Meek K M 2007 Comparative study of fibrillar collagen arrangement in the corneas of primates and other mammals *Anat. Rec.: Adv. Integr. Anat. Evol. Biol.* **290** 1542–50

[34] Morishige N, Wahlert A J, Kenney M C, Brown D J, Kawamoto K, Chikama T-i, Nishida T and Jester J V 2007 Second-harmonic imaging microscopy of normal human and keratoconus cornea *Invest. Ophthalmol. Vis. Sci.* **48** 1087–94

[35] Bueno J M, Gualda E J, Giakoumaki A, Perez-Merino P, Marcos S and Artal P 2011 Multiphoton microscopy of *ex vivo* corneas after collagen cross-linking *Invest. Ophthalmol. Vis. Sci.* **52** 5325–31

[36] Boyce B L, Jones R E, Nguyen T D and Grazier J M 2007 Stress-controlled viscoelastic tensile response of bovine cornea *J. Biomech.* **40** 2367–76

[37] Kling S, Remon L, Perez-Escudero A, Merayo-Lloves J and Marcos S 2009 Biomechanical response of normal and cross-linked porcine corneas *IOVS 2009: ARVO E-Abstract: 5477*

[38] Boyce B L, Grazier J M, Jones R E and Nguyen T D 2008 Full-field deformation of bovine cornea under constrained inflation conditions *Biomaterials* **29** 3896–904

[39] Garcia-Porta N, Fernandes P, Queiros A, Salgado-Borges J, Parafita-Mato M and Gonzalez-Meijome J M 2014 Corneal biomechanical properties in different ocular conditions and new measurement techniques *ISRN ophthalmol.* **2014** 724546

[40] Luce D A 2005 Determining *in vivo* biomechanical properties of the cornea with an ocular response analyzer *J. Cataract Refract. Surg.* **31** 156–62

[41] Fontes B M, Ambrosio R Jr, Velarde G C and Nose W 2011 Ocular response analyzer measurements in keratoconus with normal central corneal thickness compared with matched normal control eyes *J. Refract. Surg.* **27** 209–15

[42] Vinciguerra P, Albe E, Mahmoud A M, Trazza S, Hafezi F and Roberts C J 2010 Intra- and postoperative variation in ocular response analyzer parameters in keratoconic eyes after corneal cross-linking *J. Refract. Surg.* **26** 669–76

[43] Hon Y and Lam A K 2013 Corneal deformation measurement using Scheimpflug non-contact tonometry *Optom. Vis. Sci.* **90** e1–8

[44] Koprowski R 2014 Automatic method of analysis and measurement of additional parameters of corneal deformation in the Corvis tonometer *Biomed. Eng. Online* **13** 150

[45] Kling S, Bekesi N, Dorronsoro C, Pascual D and Marcos S 2014 Corneal viscoelastic properties from finite-element analysis of *in vivo* air-puff deformation *PLoS One* **9** e0104904

[46] Kling S and Marcos S 2013 Contributing factors to corneal deformation in air puff measurements *Invest. Ophthalmol. Vis. Sci.* **54** 5078–85

[47] Ariza-Gracia M A, Zurita J F, Pinero D P, Rodriguez-Matas J F and Calvo B 2015 Coupled biomechanical response of the cornea assessed by non-contact tonometry. A simulation study *PLoS One* **10** e0121486

[48] Pinsky P M and Datye D V 1991 A microstructurally-based finite element model of the incised human cornea *J. Biomech.* **24** 907–22

[49] Han Z, Sui X, Zhou D, Zhou C and Ren Q 2013 Biomechanical and refractive behaviors of keratoconic cornea based on three-dimensional anisotropic hyperelastic models *J. Refract. Surg.* **29** 282–90

[50] Glass D H, Roberts C J, Litsky A S and Weber P A 2008 A viscoelastic biomechanical model of the cornea describing the effect of viscosity and elasticity on hysteresis *Invest. Ophthalmol. Vis. Sci.* **49** 3919–26

[51] Pinsky P M, van der Heide D and Chernyak D 2005 Computational modeling of mechanical anisotropy in the cornea and sclera *J. Cataract Refract. Surg.* **31** 136–45

[52] Simonini I and Pandolfi A 2015 Customized finite element modelling of the human cornea *PLoS One* **10** e0130426

[53] Bekesi N, Kochevar I E and Marcos S 2016 Corneal biomechanical response following collagen cross-linking with rose bengal-green light and riboflavin-UVA *Invest. Ophthalmol. Vis. Sci.* **57** 992–1001

[54] Ariza-Gracia M A, Zurita J, Pinero D P, Calvo B and Rodriguez-Matas J F 2016 Automatized patient-specific methodology for numerical determination of biomechanical corneal response *Ann. Biomed. Eng.* **44** 1753–72

[55] Simonini I and Pandolfi A 2016 The influence of intraocular pressure and air jet pressure on corneal contactless tonometry tests *J. Mech. Behav. Biomed. Mater.* **58** 75–89

[56] Eliasy A, Chen K J, Vinciguerra R, Maklad O, Vinciguerra P, Ambrósio R Jr, Roberts C J and Elsheikh A 2018 *Ex-vivo* experimental validation of biomechanically-corrected intraocular pressure measurements on human eyes using the CorVis ST *Exp. Eye Res.* **175** 98–102

[57] Bekesi N, Dorronsoro C, de la Hoz A and Marcos S 2016 Material properties from air puff corneal deformation by numerical simulations on model corneas *PLoS One* **11** e0165669

[58] Roberts C J 2014 Concepts and misconceptions in corneal biomechanics *J. Cataract Refract. Surg.* **40** 862–9

[59] Kling S and Marcos S 2013 Finite-element modeling of intrastromal ring segment implantation into a hyperelastic cornea *Invest. Ophthalmol. Vis. Sci.* **54** 881–9

[60] Ariza-Gracia M A, Wu W, Malve M, Calvo B and Rodriguez Matas J F 2018 Fluid structure interaction of the non-contact tonometry test *J. Model. Ophthalmol.* **2** 75–9

[61] Nakao Y, Kiuchi Y and Okimoto S 2017 A comparison of the corrected intraocular pressure obtained by the Corvis ST and Reichert 7CR tonometers in glaucoma patients *PLoS One* **12** e0170206

[62] Alonso-Caneiro D, Karnowski K, Kaluzny B J, Kowalczyk A and Wojtkowski M 2011 Assessment of corneal dynamics with high-speed swept source optical coherence tomography combined with an air puff system *Opt. Express* **19** 14188–99

[63] Dorronsoro C, Pascual D, Perez-Merino P, Kling S and Marcos S 2012 Dynamic OCT measurement of corneal deformation by an air puff in normal and cross-linked corneas *Biomed. Opt. Express* **3** 473–87

[64] Chang E W, Kobler J B and Yun S H 2012 Subnanometer optical coherence tomographic vibrography *Opt. Lett.* **37** 3678–80

[65] Kling S, Akca I B, Chang E W, Scarcelli G, Bekesi N, Yun S-H and Marcos S 2014 Numerical model of optical coherence tomographic vibrography imaging to estimate corneal biomechanical properties *J. R. Soc. Interface* **11** 20140920

[66] Ishikawa H and Schuman J S 2004 Anterior segment imaging: ultrasound biomicroscopy *Ophthalmol. Clin. North Am.* **17** 7–20

[67] Nguyen T M, Song S, Arnal B, Huang Z, O'Donnell M and Wang R K 2014 Visualizing ultrasonically induced shear wave propagation using phase-sensitive optical coherence tomography for dynamic elastography *Opt. Lett.* **39** 838–41

[68] Touboul D, Gennisson J L, Nguyen T M, Robinet A, Roberts C J, Tanter M and Grenier N 2014 Supersonic shear wave elastography for the *in vivo* evaluation of transepithelial corneal collagen cross-linking *Invest. Ophthalmol. Vis. Sci.* **55** 1976–84

[69] Bruno L B, Fazio G and Fazio M A 2016 Custom speckle interferometer for the study of eye diseases *SPIE Newsroom*

[70] Moerman K M, Holt C A, Evans S L and Simms C K 2009 Digital image correlation and finite element modelling as a method to determine mechanical properties of human soft tissue *in viv J. Biomech.* **42** 1150–3

[71] Scarcelli G, Besner S, Pineda R, Kalout P and Yun S H 2015 *In vivo* biomechanical mapping of normal and keratoconus corneas *JAMA Ophthalmol.* **133** 480–2

[72] Liang X, Orescanin M, Toohey K S, Insana M F and Boppart S A 2009 Acoustomotive optical coherence elastography for measuring material mechanical properties *Opt. Lett.* **34** 2894–6

[73] Li C, Guan G, Reif R, Huang Z and Wang R K 2012 Determining elastic properties of skin by measuring surface waves from an impulse mechanical stimulus using phase-sensitive optical coherence tomography *J. R. Soc. Interface* **9** 831–41

[74] Li C, Guan G, Huang Z, Johnstone M and Wang R K 2012 Noncontact all-optical measurement of corneal elasticity *Opt. Lett.* **37** 1625–7

[75] Twa M D, Li J, Vantipalli S, Singh M, Aglyamov S, Emelianov S and Larin K V 2014 Spatial characterization of corneal biomechanical properties with optical coherence elastography after UV cross-linking *Biomed. Opt. Express* **5** 1419–27

Chapter 6

Ultra-high-speed Scheimpflug imaging for intraocular pressure measurements

Sven Reisdorf

Scheimpflug imaging offers an alternative method for very precise intraocular pressure (IOP) measurements as this technology has the advantage of a very high temporal and spatial resolution at the same time. The Corvis ST is the first commercially available high-speed imaging device in combination with an air-pulse tonometer. Beside highly precise IOP measurements this technology enables the measurement of corneal thickness and a detailed analysis of the biomechanical response of the cornea.

Immediately before the air pulse of the instrument starts the anterior segment of the eye is illuminated by blue light through a 9 mm slit. In the 31 ms after the beginning of the air pulse the high-speed camera captures 140 frames of the illuminated horizontal sectional plane that illustrate the deformation response of the cornea due to the air pulse. Based on these images both the intraocular pressure and the biomechanical response of the cornea can be determined.

It has been known for a long time that corneal thickness has an influence on IOP measurements. However, today it is established that the modulus of elasticity has an even higher influence than corneal thickness alone on IOP readings [1]. The Corvis ST provides an IOP reading that compensates for both corneal thickness and biomechanical properties. This ensures accurate readings even in patients with highly altered biomechanical properties such as in keratoconus or after refractive surgery.

In the following chapter the measurement principle of the Corvis ST is described and clinical studies are summarized that illustrate the advantage of this new concept for precise and accurate IOP measurements.

6.1 IOP measurements with the Corvis ST

Conventional non-contact tonometers measure intraocular pressure by determining the time of applanation of the cornea with a reflecting infrared signal. The infrared light beam is reflected on the tear film. In the original state of the cornea before the air pulse is applied the different rays are reflected into different directions and no signal is obtained by the detector. After applying the air pulse and when the cornea reaches the applanated state of the cornea the tear film acts as a flat mirror and the light beams are reflected to the photoelectric cell. At this time point a signal is obtained by the detector. The intraocular pressure is determined based on the strength of the air pulse at the time the sensor gets the signal. This method tries to mimic conventional applanation tonometry as the measured IOP is based on the amount of force that is required to applanate the cornea. However, neither the influence of corneal thickness nor the biomechanical properties are taken into consideration with this standard method. Moreover, an insufficient tear film and a bad fixation by the patient can lead to measurement errors.

Instead of using an infrared light beam the Corvis ST uses Scheimpflug technology in order to determine the time of the first applanation of the cornea.

The constant maximal pump pressure ensures that the dynamic corneal response is comparable between different individuals. The ultra-high-speed Scheimpflug camera captures 4430 frames per second. Immediately before the air pulse starts the anterior segment of the eye is illuminated by a blue light through a 9 mm slit. The blue (475 nm) slit light illuminates the anterior and posterior surfaces of the cornea and the backscattered light is captured by the camera. Based on the Scheimpflug images before the deformation of the cornea has begun, the system can also measure corneal thickness.

In the 31 ms after the start of the air pulse the high-speed camera captures 140 frames of the illuminated horizontal sectional plane. The time point of the applanation of the cornea can be accurately determined by these Scheimpflug images. As the Scheimpflug images are not dependent on the tear film the intraocular pressure measurement is also less independent of the tear film compared to conventional tonometers. Moreover, the fixation of the patient has less influence on IOP readings than in conventional NCTs. Whereas a reflected light beam will not be reflected to the sensor if the patient is not aligned correctly, in the case of the Corvis ST the time point of applanation can still be determined precisely (see figure 6.1).

These technical advantages over conventional non-contact tonometers should result in higher precision for IOP measurements.

Clinical studies have indeed demonstrated that the Corvis ST has a higher intra-observer and inter-observer repeatability and reproducibility in normal patients and glaucoma patients [2]. Very high reproducibility and good accuracy of IOP and CCT measurements were reported by Reznicek *et al* in healthy subjects and in subjects with ocular hypertension and glaucoma when compared to standard ultrasound pachymetry or Goldmann applanation tonometry [3].

6.2 Assessment of corneal biomechanical properties with the Corvis ST

Whereas the time point of the first applanation is used to get an IOP reading that is highly correlated with Goldmann applanation tonometry, all 140 images are used to

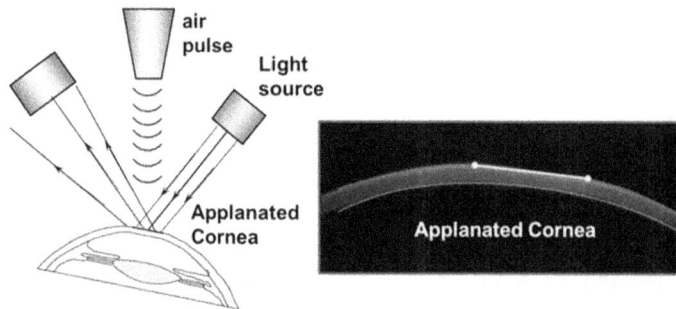

Figure 6.1. Measurement principle of a conventional NCT with an infrared light source and IR detector (a) and of the Corvis® ST equipped with a Scheimpflug camera (b). Inaccurate alignment of the cornea to the instrument will lead to a much higher measurement error in the case of a conventional NCT compared to the Corvis® ST.

depict a clear image of the corneal deformation response and to measure the biomechanical properties of the cornea [4].

The recording starts with the cornea in the initial convex state. At the beginning, the air pulse forces the cornea inwards through applanation into a concavity phase until it achieves the highest concavity.

After reaching the highest concavity, the cornea oscillates before starting to return to its original shape. Before the cornea reaches its original state, a second applanation moment occurs. Figure 6.2 shows a series of images during the deformation of the cornea as a response to the air pulse. Three events during the deformation process are of special interest:

1. The moment of the first applanation when the cornea is flat.
2. The moment of the highest concavity where the maximal deformation occurs.
3. The moment of the second applanation when the cornea is flat again before it recovers its original shape.

During all three events, different parameters describe the specific response of the measured cornea to the defined air pulse. This versatile information extracted from the direct view of the corneal deformation results in clinically relevant parameters correlated with biomechanical properties [4, 5].

One of the most important clinical parameters describing the elastic properties of the cornea is the stiffness parameter SP-A1 developed by Cynthia Roberts [4].

Corneal stiffness describes the overall rigidity of the cornea. It is defined as the applied force over displacement. In the case of the Corvis ST the force can be determined by the strength of the air pulse and the intraocular pressure whereas the displacement of the cornea can be measured directly from the Scheimpflug images. The clinical relevance of this parameter has been demonstrated for keratoconus detection, glaucoma progression and corneal crosslinking.

Whereas corneal stiffness describes the overall rigidity of the cornea, which is dependent on the intrinsic material properties and the amount of tissue, the modulus

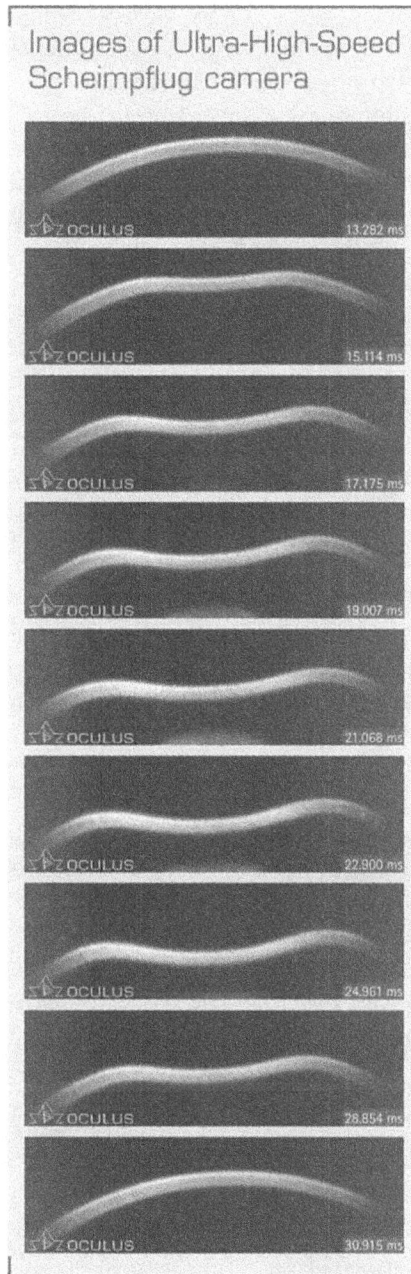

Figure 6.2. Series of Scheimpflug images showing the biomechanical response of the cornea to the air pulse.

of elasticity is only dependent on the intrinsic material properties (figure 6.3). Elsheikh and co-workers developed a method to determine the whole stress–strain curve that is relatively independent to IOP and corneal thickness. This study is currently in press. Theoretically, such an IOP and thickness independent measurement of the corneal biomechanical properties would allow a huge variety of clinical applications: from an ectasia risk assessment after refractive surgery, over patient-specific surgery planning including LASIK or corneal ring segments for keratoconus treatment.

One important application of the assessment of corneal biomechanical properties is the detection of subclinical keratoconus, especially before refractive surgery. Based on modern AI methods, combined indices—such as the Corvis biomechanical index (CBI)—have been generated that are able to detect keratoconus based on the biomechanical response of the cornea. This index has been shown to be very accurate for keratoconus detection in a big multicenter study [6]. It has also been demonstrated that one could detect cases of forme fruste keratoconus that were completely normal on corneal tomography. One step further was the integration of tomographic and biomechanical data in order to achieve the highest accuracy for the detection of forme fruste keratoconus. This concept was implemented with the development of the tomographic biomechanical index (TBI), which had the highest accuracy of all the tested methods in the original study by Ambrosio *et al* and in all following external cross-validations conducted so far [7–9].

6.3 Biomechanical corrected IOP (bIOP)

Considering the corneal biomechanical properties for an accurate IOP determination is particularly important in irregular corneas and corneas with altered stiffness. It has been known for a long time that IOP measurements in patients that have undergone refractive surgery is not accurate enough for adequate glaucoma management—either by applanation tonometry [9, 10] or by non-contact tonometry [11]. The huge biomechanical alterations in keratoconus patients or after corneal crosslinking also lead to incorrect IOP assessments with classical methods in those patients. Corneas with corneal grafts or corneas with corneal scars or edema are also difficult to measure accurately with classical methods due to the altered biomechanical behavior. Some drugs for intraocular pressure reduction such as prostaglandins are also supposed to alter the biomechanical properties of the cornea. Therefore, it is difficult to assess in these cases whether the reduced measured IOP value is just an

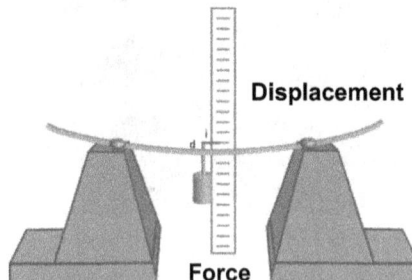

Bending Stiffness = Force / Displacement

Figure 6.3. Illustration of bending stiffness. When bending a beam the bending stiffness of the beam can be calculated by dividing the applied force by the displacement compared to the initial state.

artifact due to the altered corneal stiffness or whether the IOP reduction was sufficient for adequate glaucoma management.

Studies have shown that IOP measurement errors for only 1 mmHg can lead to incorrect clinical decisions in either glaucoma screening or glaucoma treatment [12]. Therefore, the demand is high to develop new methods that allow one to obtain accurate and precise IOP measurements in clinical situations when classical tonometry reaches its limits. It is widely accepted that the compensation for the influence of the corneal biomechanical properties on IOP measurements would be the key to IOP measurements closer to the physiological IOP in a human eye.

The measurement of corneal thickness and the biomechanical response at the same time with the Corvis ST enabled the development of an intraocular pressure value that compensates for these two factors. The so-called biomechanical corrected IOP (bIOP) was developed by Prof Ahmed Elsheikh from Liverpool. He used finite element simulations in order to analyze the dependencies of intraocular pressure measurements with corneal biomechanical properties and corneal thickness [13].

In finite element models the eye globe is distinguished into small but 'finite' elements that are described by their physical properties and which can interact with each other in the numerical simulations. Numerical simulations enable one to predict the response of the cornea to the air pulse in dependency of intraocular pressure, corneal thickness and corneal biomechanics. This 3D finite element model has been validated in experimental studies.

By running several simulations the relationships between intraocular pressure measurements (without correction), corneal thickness and biomechanical properties can be investigated in detail with this approach.

Based on this knowledge, the analyzed dependencies can be used to create an analytical formula that compensates for both corneal thickness and corneal biomechanical properties. This equation is directly implemented in the device so that this pressure value is readily presented to the user. In a first step the accuracy of this biomechanical corrected IOP (bIOP) was validated in human cadaver eye studies [14]. Cadaver eyes have the advantage that the pressure inside the eye can be accurately adjusted and is therefore known. The measured IOP with the tonometer can be compared to the adjusted intraocular pressure inside the cadaver eyes. Indeed, the study proved that the bIOP was very closely related to the physiological IOP inside the cadaver eyes and much more accurate than the conventional IOP readings (figure 6.4) [14].

6.4 Simulation-based versus empirically-derived correction formulas

This approach has several advantages compared to conventional methods to develop IOP correction formulas. Conventional formulas were developed solely on empirical data. The amount of correction is just based on the relatively weak correlation of the derived IOP values with corneal thickness in a specific patient group. Due to the relatively weak correlation outliers can have a significant influence on the regression analysis. Moreover, in a different population these kinds of analyses could lead to slightly different equations; and most importantly, if a patient

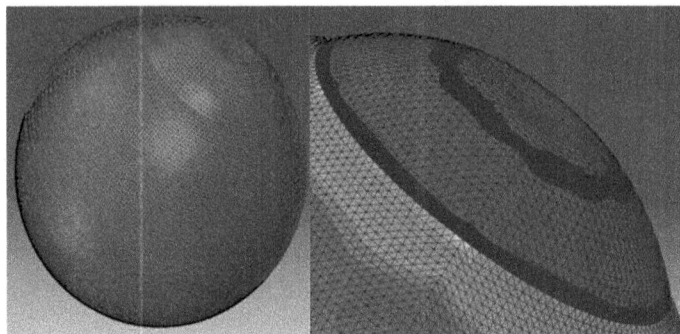

Figure 6.4. Finite element models of the whole globe, the scleral region and the cornea that are used to develop the biomechanical corrected IOP (bIOP). Reproduced from [13]. Reprinted by permission of the publisher (Taylor & Francis Ltd, http://www.tandfonline.com).

doesn't fall into this group of patients the equations can lead to results that are completely wrong. Of course, the classical empirically-derived correction formulas can only correct for corneal thickness and do not consider biomechanical properties at all.

For example, patients with edema have very thick but soft corneas. Applying an empirically-derived thickness-based correction formula would lead to a down correction of the measured IOPs. However, the physiological IOP would be higher than the measured IOP due the soft tissue. In numerical simulations these dependencies are considered, which allows a patient-specific IOP correction. The extent of validity of the derived equation is also known from the simulations. Therefore, patient-group specific equations are implemented in the Corvis ST that have been proven to be accurate for exactly this group of patients with a similar biomechanical behavior as the specific patient.

In the following section clinical applications are shown that demonstrate the value of the bIOP for accurate measurements even with highly altered biomechanical properties.

6.5 Accuracy of bIOP in patients after refractive surgery

One of the most challenging clinical situations for accurate IOP measurements is in patients who have undergone corneal refractive surgery in the past. Many studies have demonstrated clinically relevant changes in IOP compared to the pre-operative state. Whereas the 'true' IOP very likely remains the same as the pre-operatively, these measurement changes reflect measurement errors as the basic assumptions of IOP measurements are violated. The resulting underestimation of IOP in these cases can lead to missed diagnosis of POAG or OHT.

In particular, GAT measurements are known to underestimate IOP a lot after refractive surgery—in some studies by 5 mmHg as an average [15]. Despite the high mean average the standard deviation is even higher and the measurement error is not accurately predictable. The reason for the failure of applanation tonometry after

LVC is that the method relies on assumptions regarding the shape, thickness and biomechanics of the cornea, which are all invalid after any kind of laser vision correction.

In a paper published by Lee *et al* the newly developed bIOP was tested in patients before and after t-PRK and LASIK [16]. The study showed that there was no significant change in bIOP in both procedures compared to the pre-operative state (129 patients were included in this study). The Corvis ST measurements were performed before the procedure and 6 month's post-operatively.

In a second paper the same group analyzed patients that underwent t-PRK combined with accelerated crosslinking before and after the procedure and compared these patients with patients that only underwent t-PRK without crosslinking [17]. Once more, no significant differences in the bIOP occurred post-operatively relative to the pre-operative bIOP in either group.

Chen *et al* published a further paper where they compared the bIOP before and after SMILE in addition to LASIK patients [18]. Once more, the study demonstrated that in both procedures the bIOP was stable before and after the procedure.

Beside the Corvis ST, the dynamic contour tonometer is also supposed to be relatively independent of corneal biomechanical properties. The ocular response analyzer provides an IOP value—the corneal compensated IOP (IOPcc)—which is also supposed to be less dependent on corneal biomechanical properties. In a recently submitted paper, the effectiveness of the Goldmann applanation tonometer, the dynamic contour tonometer, the ocular response analyzer and the Corvis ST in measuring IOP following FS-LASIK were compared [19]. The main outcome of the study was that '*the biomechanically-corrected bIOP from the Corvis provides post-surgery measurements that were in closest agreement with those obtained before FS-LASIK, and bIOP was followed closely by DCT-IOP. GAT-IOP, ORA-IOPcc, ORA-IOPg and CVS-IOP appeared to be more influenced by the effects of corneal biomechanics changes caused by LASIK*'. Furthermore, bIOP was not correlated with CCT, Km and age either before or after FS-LASIK.

Despite analyzing the stability of bIOP before and after refractive surgery, cases of pressure-induced stromal keratopathy were reported where the Goldmann applanation tonometer measurements completely failed with a measurement of low pressure, while the Corvis ST enabled a more accurate measurement of high pressure [20]. This case report illustrates that Corvis ST measurements can provide more accurate IOP readings after LASIK even in the rare case of a pressure-induced stromal keratopathy. The Scheimpflug images are extremely helpful for the visualization of IOP changes in these cases.

6.6 bIOP measurements in patients with keratoconus

Keratoconus patients have thinner, steeper and softer corneas than healthy patients and therefore do not comply with most of the assumptions of applanation tonometry. This makes keratoconus one of the most challenging clinical situations in which to obtain accurate IOP readings [21, 22]. The modulus of elasticity is changed by such an amount that even the bIOP equation that is valid for healthy patients cannot be applied anymore. Therefore, Prof Elsheikh and his team developed another equation that can correct the influence for patients with elastic modulus and corneal shape that are completely altered. The correct formula is automatically chosen based on the biomechanical properties of the cornea so that the user doesn't recognize that different analytical equations were used. This shows again the strength of numerical simulations as the extent of validity for each equation is exactly known.

The accuracy of the bIOP on keratoconus patients has been tested in a study submitted by Chen *et al*. In this multicenter study 722 eyes of 722 participants (369 healthy, 353 keratoconus) were tested [23]. The keratoconus eyes of this study had wide variations in thickness, geometry and tissue stiffness. The main outcome of the study was the absence of a significant difference ($p < 0.05$) on bIOP readings on healthy and keratoconus patients whereas the uncorrected IOP readings were significantly different. Furthermore, the bIOP was also less influenced by both corneal thickness and age.

Hugo *et al* analyzed IOP readings obtained by the Corvis ST in corneas after keratoplasty and compared them with normal controls [24]. In this study, IOP reading in eyes after keratoplasty were also very comparable to the normal controls despite significant changes in biomechanical properties.

As a main conclusion, current studies also proved that the Corvis ST bIOP can provide valid IOP readings in extremely altered biomechanical properties and corneal shape such as post-corneal refractive surgery, in keratoconus patients or after keratoplasty.

6.7 Ongoing Corvis ST studies in glaucoma

High precision tonometry is the first requirement for adequate glaucoma management. Very high reproducibility and good accuracy of IOP measurements were reported by Reznicek *et al* in healthy subjects and in subjects with ocular hypertension and glaucoma when compared to Goldmann applanation tonometry [3].

Despite high precision, it is of major interest whether a new IOP reading can be more effective to distinguish between ocular hypertension, high and normal tension, primary open glaucoma and normal eyes. The ultimate goal would be to identify IOP readings that are more accurate to predict glaucoma progression. In a currently submitted study, Vinciguerra *et al* compared bIOP readings with Goldman applanation tonometry in all four groups [25]. Interestingly, bIOP was lower in patients with ocular hypertension compared to applanation tonometry indicating less overestimation in this patient group.

More studies are currently ongoing analyzing the bIOP in different glaucoma populations. One important clinical question is the influence of corneal biomechanical properties on IOP measurements in patients treated with prostaglandins. A significant amount of the measured pressure reduction after prostaglandin treatment is very likely a measurement artifact due to alterations of biomechanical properties due to medication. Potentially, the bIOP offers a new method to determine more accurately the reduction of intraocular pressure after prostaglandin treatment.

Despite IOP measurements, the biomechanical properties measured with the Corvis ST have been demonstrated to be independent risk factors for glaucoma [26]. In a longitudinal study, Craig and co-workers have shown that the stiffness at the baseline measurement is a predictor of glaucoma progression (submitted). Further studies on glaucoma progression have been performed by Asaoka and co-workers [27, 28].

Moreover, it has been shown is several studies that in normal tension glaucoma the cornea is more deformable compared to healthy corneas of patients with the same age. Currently, a software is being developed including normative data from healthy patients and patients with normal tension glaucoma that will aid in screening for normal tension glaucoma patients at an early stage.

6.8 Summary

The Corvis ST provides a new approach to accurately measure the intraocular pressure, corneal thickness and the biomechanical properties of the cornea. Based on the combination of a high-speed Scheimpflug camera with an air-pulse tonometer, the device provides highly precise IOP readings with similar or even higher reproducibility than the current gold standard—the Goldman applanation tonometry. The assessment of corneal biomechanical properties enabled the development of the bIOP, which has been demonstrated to be more accurate than other IOP readings in situations with highly altered biomechanical situations and corneal shape such as post-refractive surgery, keratoconus or keratoplasty. Furthermore, it has been shown to be less affected by influencing factors such as corneal thickness, age, shape and corneal stiffness. Recent studies indicate that the bIOP might show less overestimation of intraocular pressure compared to applanation tonometry in patients with ocular hypertension.

Beside the high value as a tonometer, the measurement of biomechanical properties with the Corvis ST is of high clinical relevance. Improved sensitivity and specificity for the screening of forme fruste keratoconus has been demonstrated when combining tomographic data with the biomechanical data of the Corvis ST. Other studies have proven that corneal biomechanical properties of the cornea can be used as independent risk factors for glaucoma.

Currently, a new analysis software based on machine learning is being developed including normative data from healthy patients and patients with normal tension glaucoma that will aid in screening for normal tension glaucoma patients at an early stage.

References

[1] Liu J and Roberts C J 2005 Influence of corneal biomechanical properties on intraocular pressure measurement: quantitative analysis *J. Cataract Refract. Surg.* **31** 146–55

[2] Hong J, Xu J, Wei A, Deng S X, Cui X, Yu X and Sun X 2013 A new tonometer—the Corvis ST tonometer: clinical comparison with noncontact and Goldmann applanation tonometers *Invest. Ophthalmol. Vis. Sci.* **54** 659–65

[3] Reznicek L, Muth D, Kampik A, Neubauer A S and Hirneiss C 2013 Evaluation of a novel Scheimpflug-based non-contact tonometer in healthy subjects and patients with ocular hypertension and glaucoma *Br. J. Ophthalmol.* **97** 1410–4

[4] Roberts C J, Mahmoud A M, Bons J P, Hossain A, Elsheikh A, Vinciguerra R, Vinciguerra P and Ambrósio R Jr 2017 Introduction of two novel stiffness parameters and interpretation of air puff induced biomechanical deformation response parameters with a dynamic Scheimpflug analyzer *J. Refract. Surg.* **33** 266–73

[5] Vinciguerra R, Elsheikh A, Roberts C J, Kang D S Y, Lopes B T, Morenghi E, Azzolini C and Vinciguerra P 2016 Influence of pachymetry and intraocular pressure on dynamic corneal response parameters in healthy patients *J. Refract. Surg.* **32** 550–61

[6] Vinciguerra R, Ambrósio R Jr, Elsheikh A, Roberts C J, Lopes B, Morenghi E, Azzolini C and Vinciguerra P 2016 Detection of keratoconus with a new biomechanical index *J. Refract. Surg.* **32** 803–10

[7] Ambrósio R Jr, Lopes B T, Faria-Correia F, Salomão M Q, Bühren J, Roberts C J, Elsheikh A, Vinciguerra R and Vinciguerra P 2017 Integration of Scheimpflug-l based corneal tomographic and biomechanical assessments for enhancing ectasia detection *J. Refract. Surg.* **33** 266–73

[8] Chan T C Y, Wang Y M, Yu M and Jhanji V 2018 Comparison of corneal tomography and a new combined tomographic biomechanical index in subclinical keratoconus *J. Refract. Surg.* **34** 616–21

[9] Chatterjee A, Shah S, Bessant D A, Naroo S A and Doyle S J 1997 Reduction in intraocular pressure after excimer laser photorefractive keratectomy: correlation with pretreatment myopia *Ophthalmology* **104** 355–9

[10] Mardelli P G, Piebenga L W, Whitacre M M and Siegmund K D 1997 The effect of excimer laser photorefractive keratectomy on intraocular pressure measurements using the Goldmann applanation tonometer *Ophthalmology* **104** 945–9

[11] El Danasoury M A, El Maghraby A and Coorpender S J 2001 Change in intraocular pressure in myopic eyes measured with contact and non-contact tonometers after laser *in situ* keratomileusis *J. Refract. Surg.* **17** 97–104

[12] Turner M J, Graham S L, Avolio A P and Mitchell P 2013 Potential effects of systematic errors in intraocular pressure measurements on screening for ocular hypertension *Eye* **27** 502–6

[13] Joda A A, Shervin M M, Kook D and Elsheikh A 2016 Development and validation of a correction equation for Corvis tonometry *Comput. Methods Biomech. Biomed. Eng.* **19** 943–53

[14] Eliasy A, Chen K-J, Vinciguerra R, Maklad O, Vinciguerra P, Ambrósio R Jr, Roberts C J and Elsheikh A 2018 *Ex-vivo* experimental validation of biomechanically-corrected intra-ocular pressure measurements on human eyes using the CorVis ST *Exp. Eye Res.* **175** 98–102

[15] Fan Q, Zhang J, Zheng L, Feng H and Wang H 2012 Intraocular pressure change after myopic laser *in situ* keratomileusis as measured on the central and peripheral cornea. Clinical and experimental *Ophthalmology* **95** 421–6

[16] Lee H, Roberts C J, Kim T I, Ambrósio R Jr, Elsheikh A and Kang D S Y 2017 Changes in biomechanically corrected intraocular pressure and dynamic corneal response parameters before and after transepithelial photorefractive keratectomy and femtosecond laser-assisted laser *in situ* keratomileusis *J. Cataract Refract. Surg.* **43** 1495–503

[17] Lee H, Roberts C J, Ambrósio R Jr, Elsheikh A, Kang D S Y and Kim T I 2017 Effect of accelerated corneal crosslinking combined with transepithelial photorefractive keratectomy

on dynamic corneal response parameters and biomechanically corrected intraocular pressure measured with a dynamic Scheimpflug analyzer in healthy myopic patients *J. Cataract Refract. Surg.* **43** 937–45

[18] Chen K J, Joda A, Vinciguerra R, Eliasy A, Sefat S M M, Kook D, Geraghty B, Roberts C J and Elsheikh A 2018 Clinical evaluation of a new correction algorithm for dynamic Scheimpflug analyzer tonometry before and after laser *in situ* keratomileusis and small-incision lenticule extraction *J. Cataract Refract. Surg.* **44** 581–8

[19] FangJun B *et al* 2019 Effectiveness of the Goldmann applanation tonometer, the dynamic contour tonometer, the ocular response analyzer and the Corvis ST in measuring intraocular pressure following FS-LASIK (in preparation)

[20] Faria-Correia F, Ramos I, Valbon B, Luz A, Roberts C J and Ambrósio R 2013 Scheimpflug-based tomography and biomechanical assessment in pressure-induced stromal keratopathy *J. Refract. Surg.* **29** 356–8

[21] Shah S, Laiquzzaman M, Bhojwani R, Mantry S and Cunliffe I 2007 Assessment of the biomechanical properties of the cornea with the ocular response analyzer in normal and keratoconic eyes *Invest. Ophthalmol. Vis. Sci.* **48** 3026–31

[22] Altinkaynak H, Kocasarac C, Dundar H, Sayin N, Kara N, Bozkurt E and Duru N 2016 Which tonometry in eyes with keratoconus? *Eye* **30** 431–7

[23] Chen K-J, Eliasy A, Vinciguerra R, Ahmed A, Vinciguerra P, Ambrósio R Jr, Roberts C J and Elsheikh A 2019 Development and validation of a new intraocular pressure estimate for patients with keratoconus (in preparation)

[24] Hugo J, Granget E, Ho Wang Yin G, Sampo M and Hoffart L 2018 Intraocular pressure measurements and corneal biomechanical properties using a dynamic Scheimpflug analyzer, after several keratoplasty techniques, versus normal eyes *J. Fr. Ophtalmol.* **41** 30–8

[25] Vinciguerra R *et al* 2019 Corneal biomechanics and biomechanically-corrected intraocular pressure measurements in patients with primary open angle glaucoma, ocular hypertension and controls (in preparation)

[26] Li B B, Cai Y, Pan Y Z, Li M, Qiao R H, Fang Y and Tian T 2017 Corneal biomechanical parameters and asymmetric visual field damage in patients with untreated normal tension glaucoma *Chin. Med. J.* **130** 334–9

[27] Matsuura M, Hirasawa K, Murata H, Nakakura S, Kiuchi Y and Asaoka R 2017 Using Corvis ST tonometry to assess glaucoma progression *PLoS One* **12** e0176380

[28] Matsuura M, Hirasawa K, Murata H, Nakakura S, Kiuchi Y and Asaoka R 2017 The usefulness of Corvis ST tonometry and the ocular response analyzer to assess the progression of glaucoma *Sci. Rep.* **7** 40798

www.ingramcontent.com/pod-product-compliance
Lightning Source LLC
Chambersburg PA
CBHW082110210326
41599CB00033B/6653